AF805053

ONE WEEK LOAN

25 0291187 4

THE INTERNET
IN THE ARAB WORLD

Digital Formations

Steve Jones
General Editor

Vol. 43

PETER LANG
New York • Washington, D.C./Baltimore • Bern
Frankfurt am Main • Berlin • Brussels • Vienna • Oxford

Rasha A. Abdulla

THE INTERNET IN THE ARAB WORLD

Egypt and Beyond

PETER LANG
New York • Washington, D.C./Baltimore • Bern
Frankfurt am Main • Berlin • Brussels • Vienna • Oxford

W302.23 ABD

Library of Congress Cataloging-in-Publication Data

Abdulla, Rasha A.
The Internet in the Arab world: Egypt and beyond / Rasha A. Abdulla.
p. cm. — (Digital formations; v. 43)
Includes bibliographical references and index.
1. Information technology—Arab countries. 2. Internet users—Egypt.
3. Internet—Religious aspects—Islam. 4. Censorship—Arab countries. I. Title.
HN786.Z9I5615 303.48'3309174927—dc22 2006022713
ISBN 978-0-8204-8673-4
ISSN 1526-3169

Bibliographic information published by **Die Deutsche Bibliothek**.
Die Deutsche Bibliothek lists this publication in the "Deutsche Nationalbibliografie"; detailed bibliographic data is available on the Internet at http://dnb.ddb.de/.

Cover design by Joni Holst

The paper in this book meets the guidelines for permanence and durability of the Committee on Production Guidelines for Book Longevity of the Council of Library Resources.

© 2007 Peter Lang Publishing, Inc., New York
29 Broadway, 18th floor, New York, NY 10006
www.peterlang.com

All rights reserved.
Reprint or reproduction, even partially, in all forms such as microfilm, xerography, microfiche, microcard, and offset strictly prohibited.

Printed in Germany

To my parents . . .
And to those who live and let live, and understand that different is good.

CONTENTS

FOREWORD TO THE ARABIC EDITION

Hussein Amin

This important volume is considered an invitation for all readers to elevate their basic knowledge about the Internet in the Arab world in general, and specifically about the uses, effects, and future of this medium in the Arab region. The author is gaining increasing recognition in the field and her contributions to the study of Internet in the region are invaluable. This book is based on the author's years of scientific observation and research about information and communication technologies and new media, particularly the Internet, both in Egypt and in the United States.

The book is a simplified effort to understand a complex and pertinent issue. It provides essential information and knowledge about the development of the Internet in the Arab world and its effect on shaping the Arab mind. The book presents an important scientific study and documents a comprehensive review of literature in an easy to understand manner, making it a valuable and enjoyable read to ICT specialists and non-specialists, communication scholars and students, and anyone with an interest in the Arab world.

The author starts by presenting an overview of the media in the Arab world in general and the new directions that each new medium has taken us into. She then details the growth of the Internet in the Arab world, and provides a comprehensive presentation of the uses and effects of this new medium on the region.

The book aims to present an analysis of the Internet as a unique communication medium in the Arab world. One main factor is the potential of the Internet to provide a new tool of expression for nongovernmental and nonofficial civic groups and the ways in which these voices can turn the Arab community into a more liberal and development-oriented community. Another important discussion is whether or not the Internet can help in strengthening democracy in the Arab world and in creating an informed Arab public opinion. The viewpoints presented in this book have pointed to the difficulty of developing a truly engaged, politically active Arab community without the Internet. The author's view is that enhancing Arabs' living conditions and their understanding of each others' and of world cultures will not happen without the Internet.

Despite the fact that most Arab countries share the same language, culture, religion, and history, there are core differences among Arab audiences regarding political and cultural ideologies. The book discusses how Arabs can all experience one world—maybe for the first time in their long history—without borders or boundaries blocking the flow of information and events and communicate with each other without censorship, through an instant communication channel, which breaks the boundaries of time and place. The Internet has presented people in the region with the opportunity to engage in conversations, discussions, and debates about issues which were previously considered taboo such as religion, women's rights, and Arab governments, among others.

The Internet has also given Arabs an opportunity to freely express themselves, thus enhancing and stimulating social and intellectual interactions. It has presented a cultural meeting point for communication among different nationalities and individual voices, which form the basis for new and free virtual political gatherings and institutions. The Internet also gives the Arab world the opportunity to present itself to an international audience and to share its history and culture on the international level.

The author points out the challenges facing Internet use in the Arab world, especially the cultural fears in some Arab communities regarding the potential effects of the Internet. The Arab world is in great need to open up to foreign cultures and reach a level of balance between what it receives from the outside world and what it shares with it in terms of national and cultural identity. It is hard nowadays to see the world from one perspective and to only react in a defensive manner to the outside world.

This book includes a number of valuable views and answers regarding general questions about how Arabs are using the Internet, and what they are likely to do with it for their future. Will Arabs use the Internet as a rich cultural and political source, as an educational tool, and a global source of information? This will only enrich their cultural and political agendas and enliven their intellectual heritage.

I believe this book is an excellent addition to the library of those interested in the Arab region, especially for its calls for challenging discussions of the use of ICT in the region.

Hussein Amin, Ph.D.
Professor and Chair, Journalism and Mass Communication
The American University in Cairo

FOREWORD TO THE ENGLISH EDITION

Douglas A. Boyd

Most communication scholars believe that the study of communication patterns is among the best ways to learn about a region. Scholarly communication studies have served to help the world understand a region of increasing interest since the Gulf War, 9/11, and the 2003 invasion of Iraq.

One of the more promising developments in the past decade of Middle East media research is the emergence of young, well-trained scholars who reside in the region and produce important work in both Arabic and English. Heretofore, much of the research published in Arabic did not reach beyond the Middle East, and thus we were largely unaware of it.

Rasha Abdulla's book on Arab media, focusing on the Internet, is precisely the type of quality work that is needed to help all who are interested in Middle East politics, culture, and economics understand a phenomenon that has been well studied in the West, but not extensively researched in Arab countries. This book is a strong beginning. It addresses the need to understand how the Internet functions in the Arab world.

Dr. Abdulla first introduces the reader to Arab media and telecommunications policies. Egypt, the author's home, provides the basis for the discussion that then moves to other portions of the Arab world. The publication especially helps redress the lack of theory-based Arab communication scholarship.

After reading the book, readers will know why Rasha Abdulla is one of the Arab world's most promising young media scholars.

Douglas A. Boyd, Ph.D.
Professor, Department of Communication,
School of Journalism and Telecommunications,
Patterson School of Diplomacy and International Commerce
Chief of Staff, Office of the President, University of Kentucky
Author of Broadcasting in the Arab World:
A Survey of the Electronic Media in the Middle East.

PREFACE

I started getting interested in communication technologies some years before the Internet became popular. In particular, my interest began in 1989 while still a sophomore at the Journalism and Mass Communication Department of the American University in Cairo (AUC), where I now teach as an Assistant Professor. In that year, I joined the *Caravan*, the student newspaper of AUC, and it so happened that I was assigned to cover a story on the introduction of BITnet to AUC. I didn't have much information to fall back onto. I remember my editor saying, "They're introducing something called BITnet to AUC and they're making a big fuss about it. Go find out what it is."

And so I did. I carried my little notebook, my excitement to be a "real" journalist, and my curiosity and set out to figure out exactly what BITnet was. After listening to the introductory remarks, I realized that BITnet stood for "Because It's Time Network," and that apparently, it can do very cool things. OK, great, but the question remained: Because It's Time to do what?

With my budding journalistic skills striving hard to bloom at age 16, I went around interviewing people in an attempt to understand. They told me that BITnet was a communication network through which I could communicate with people all around the world by sending and receiving instant computer messages. It was just like mailing a letter, but much faster and a lot more efficient. It's what is called electronic mail, or e-mail.

WOW! That got me hooked. At a time when mailing a letter from Egypt to the United States carried with it a considerable risk of at best reaching its destination in 3–8 weeks and at worst never getting there, this sounded unbelievable. I was heartbroken however to find out that those BITnet accounts would only be available to AUC faculty at that time and not to students. There were just not enough resources. Professors, it was determined, were the ones who stood to benefit the most through sharing academic materials with counterparts all over the world. Despite this setback to my personal agenda, I didn't lose interest in what seemed to me to be an amazing application of telecommunications technology. I "volunteered" at the *Caravan* to cover all events related to BITnet and used my journalistic leverage to gain access to those lectures, presentations, and meetings to satisfy my curiosity about how this resource was being used.

My interest in how such technologies are used never faded. Later on in 1993, the Internet was introduced to Egypt, and I was among the very first to acquire an account, learn to use it, and actually use it. I remember my excitement whenever a new Internet application or feature would surface. First it was all text, and it was mostly e-mail. Then it was Gopher, the magic application through which you could find possibly anything although you really needed to know how and where to look. Then Mosaic and the introduction of color and graphics. And finally, the World Wide Web as we know it today: a vast territory of information about anything you ever dreamed of, easily accessible through a variety of search engines that actually tell you where to go to find the information. Fascinating!

To give you an idea of how pioneering my humble knowledge was at the time, I still keep in my personal files a "historic" document from the year 1995. This memo, addressed to me from the American country director of the American NGO I worked for back then awarded me a L.E. 500 (around $150, and 25% of my lucrative personal salary at the time) for "all (my) hard work in diligently pursuing and securing an Internet account through (my) personal contacts" for the NGO. The memo continued, "Through you (me, that is), our Egypt office became the first field office (out of 13 worldwide) to enter the Internet world. This account will greatly contribute to the enhancement of our services." The memo then sang my praises, "We all share our pride in you and appreciate your remarkable contribution to this organization. Thanks again for your hard work." This was how hard and how rare it was to acquire an Internet account back then. You'll see throughout this book how things have changed.

The Internet took on a whole new meaning for me when in 1999 I got a scholarship to do my Ph.D. in the United States. Although I had traveled abroad frequently before, I had never lived alone abroad for extended periods of time. In the Arab world, we do not move out of our parents' house in our late teens like it is the case in the Western world. (This applies whether you're male or female). Instead, we live with our families, usually until we get married and move to a family house of our own. As an only child of an Egyptian family, and female nonetheless, the over-attachment between my parents and myself began to haunt me. And in my case, you need to keep in mind that my parents, whom I love dearly, are not exactly the least worrying of people. One of the arguments I used to get all three of us (myself included) comfortable with the idea of my leaving to the United States was to tell my parents I would e-mail them once or twice every day, so hopefully it would be a bit easier for all of us. It actually worked. My parents told me on one of my summer vacations in Cairo that they feel closer to my life through the frequent e-mail communications than they felt when I was living with them in the same house. It is amazing what e-mail can do.

Perhaps the most invaluable personal moment the Internet presented to my family was when it enabled my parents to watch my doctoral graduation ceremony streamed live from the University of Miami through their computer screen in Cairo, Egypt. They followed the event moment by moment, until they called me on my cell phone right after I stepped out of the UM Convocation Center where the commencement was held. If that does not make the world a global village, I don't know what does.

I am aware however, of some potentially negative effects or uses of the Internet as well. For example, while working on my doctorate in the United States, I noticed that a few Arab students used to spend hours at a time chatting with their families and friends back home. This does not only waste their days in the United States and take away from the time they're supposed to be studying, but just as importantly if not even more importantly, it prevents them from getting exposed to the different culture they are living in. It stands in the way of their going out and socializing with American or other international students, and therefore in the way of understanding these foreign cultures. They end up missing out on a major aspect of their journey to the United States. They do not get exposed to Americans or foreigners, and therefore do not learn about their ways of thinking or their points of view. Instead they live in their virtual connections to their home countries, in what they think is "protecting" them from the need to have foreign friends or experience foreign cultures.

I am also familiar with the arguments that were loud a few years ago in the Arab world pertaining to the potential of the Internet to provide access to children and youths to pornography or other materials considered "inappropriate" by their parents or by society. However, my conviction is that the Internet, like any other medium, can be a tool of positive or negative potential depending on how we, the audience, choose to use it. I do believe though, as I'll try to demonstrate in this book, that the Internet offers the Arab world tremendous potential for development and reform.

It's for all these reasons that when it was time for me to conduct my doctoral dissertation, I knew what I wanted to research. In fact, I had been gathering information on it for about four years. I wanted to know how Arab students in Egypt use the Internet, and what potential this medium brings to Egypt and the Arab world. How do students perceive this new medium? What do they think of it? Do they perceive it as a threat or an opportunity? How much time do they spend on it? How do they use it? For what purposes? Can the Internet be a tool for free expression, otherwise a rare commodity, in the Arab world? Can the Internet be a tool for enhancing civic society, and for bringing about more democracy? Ultimately, can the Internet help Egypt and the Arab world achieve higher levels of development and better living standards? Can the Internet provide for better relations and understanding between the West and the Arab world? And if so, then how?

This book attempts to answer these questions. It is the outcome of my doctoral dissertation (2003) at the University of Miami, Coral Gables, Florida, as well as of later research and of years and years of curiosity and observation. I would like to thank a few people without whom this work would not have been possible: My dissertation supervisor, Dr. Gonzalo Soruco, and readers Drs. Bruce Garrison, Diane Millette, and Eugene Provenzo, all of the University of Miami; and Damon Zucca, Steve Jones, and Sophie Appel of Peter Lang Publishing, Inc. I would of course like to thank my parents for so many things, not the least of which is bearing with my absence for four and a half years while trying to learn and answer some of these important questions, but also for their continuous encouragement and prayers, and their belief in me and my humble abilities. I would like to thank my friends, and they know who they are, for always standing by me and taking pride in what I accomplish. And of course before all, I thank God for blessing me with the passion to learn and with so many good people in my life.

I hope you find this book useful and enjoyable. If you have any comments or questions, you can reach me at rasha@aucegypt.edu or through my Web page at http://www.aucegypt.edu/faculty/rabdulla or www.rashaabdulla.com.

Thank you.

Rasha A. Abdulla, Ph.D.
Assistant Professor
Journalism and Mass Communication
The American University in Cairo
September 2006

ABBREVIATIONS

AFESD	Arab Fund for Economic and Social Development
ANOVA	Analysis of Variance
ANRT	Agence Nationale de Règlementation des Télécommunications (National Telecommunications Regulatory Agency)
AOL	America Online
ARENTO	Arab Republic of Egypt National Telecommunication Organization
ARPA	Advanced Research Projects Agency
ATI	Agence Tunisienne d'Internet (Tunisian Internet Agency)
ATM	Asynchronous Transfer Mode
AUC	American University in Cairo
BBC	British Broadcasting Corporation
BITnet	Because It's Time Network
CERN	European Organization for Nuclear Research
CMC	Computer-Mediated Communication
CNN	Cable News Network
CSPs	Certificate Service Providers
DBS	Direct Broadcast Satellite

DIC	Dubai Internet City
DNS	Domain Name Server
EARN	European Academic and Research Network
EIM	Emirates Internet and Multimedia
ESC	Egyptian Satellite Channel
ESV	Egypt Smart Village
EUN	Egyptian Universities Network
FEMAS	Federation of Maranao Associations in Saudi
FLAG	Fibre Optic Link Around the Globe
GITR	Global Information Technology Report
GSM	Global System for Mobile Communications
HRW	Human Rights Watch
HTML	HyperText Markup Language
ICANN	Internet Corporation for Assigned Names and Numbers
ICT	Information and Communication Technology
IDSC	Information and Decision Support Center
ISDN	Integrated Services Digital Network
ISPs	Internet Service Providers
ISU	Saudi Internet Services Unit
IT	Information Technology
ITIDA	Information Technology Industry Development Agency
ITU	International Telecommunication Union
JTC	Jordan Telecommunications Corporation
KACST	King Abdul-Aziz City for Science and Technology
MBC	Middle East Broadcasting Corporation
MCIT	Ministry of Communications and Information Technology
MOHP	Ministry of Health and Population
MPC	Media Production City
MTC	Mobile Telecommunications Company
NCSA	National Center for Supercomputing Applications
NGOs	Non-Governmental Organizations
NIC	National Information Center
NMTC	National Mobile Telecommunications Company
NSF	National Science Foundation
PDAs	Personal Data Assistants
RCDs	Remote Control Devices
RITSEC	Regional Information Technology and Software Engineering Center

RSF	Reporters Sans Frontieres
STC	Saudi Telecommunications Company
STE	Syrian Telecommunications Establishment
TACCs	Technology Access Community Centers
TCP/IP	Transmission Control Protocol/Internet Protocol
UAE	United Arab Emirates
UNDP	United Nations Development Programme
URLs	Uniform Resource Locators
USIS	United States Information Service
VCRs	Video Cassette Recorders
VIF	Variance-Inflation Factor
VOA	Voice of America
W3C	World Wide Web Consortium
WEF	World Economic Forum
WSIS	World Summit on the Information Society
WWW	World Wide Web

· 1 ·

MEDIA TECHNOLOGIES IN EGYPT AND THE ARAB WORLD

The advances in modern communication and information technologies are so rapid that sometimes we cannot keep up. The implications of such advances are far reaching. For one thing, communication, in the sense of delivering a message to another person or a group of people, has become much easier and much less expensive. Another important implication is that the world is expected to become more and more of a global village. As early as 1962, Marshall McLuhan argued that "the new electronic interdependence recreates the world in the image of a global village" (p. 31). At the time, he was speaking of the effects of "new" media like radio and television, which he argued were to bring a sense of community and equal access to information to the world. Two years later, he wrote, "After more than a century of electric technology, we have extended our central nervous system in a global embrace, abolishing both space and time as far as our planet is concerned" (1964, p. ix). Some scholars now believe that with the introduction of the Internet, McLuhan's image of a global village is becoming more of a reality than a metaphor.

There is no question that the introduction of new technologies, from telephony to television to the Internet, has made it easier to overcome physical distances and time differences, and made it more convenient for people to communicate with each other and/or through the mass media. However, some questions remain: How are we using these technologies? What functions have these technologies served us, and how are we using them to fulfill our needs? Are we using these technologies to their full potential to make our world a better place? Have these technologies brought the world any closer or widened an already existing cultural gap? Can these technologies help us get acquainted with new cultures and new people? Can we use these technologies to help us learn, understand, become more socially fulfilled, become better informed and more entertained? Can the new technologies help us appreciate, and tolerate "other" peoples and "other" cultures?

I was studying for my doctorate in the United States when the tragic events of September 11, 2001, took place. Aside from the gravity of this most tragic situation, the lack of understanding between Western cultures, particularly the United States in this case, and the Arab world became immediately clear to me. People on both sides made statements that revealed a serious lack of knowledge and understanding about the "other." Can our use of technology, particularly the Internet, help us bridge that gap? Can Arabs use the Internet to teach the world about Arab culture, religions, and what Islam and Muslims stand for? And what other functions can the Internet provide for Arab users?

To answer these questions, we need to understand how people in the Arab world use the Internet, and the potential effects that this medium can have on the Middle East and the Arab world. Studies, carried out mostly in the United States and Europe, have begun to reveal patterns of Internet use in the West. Little is known, however, about how the Arab world uses technology, particularly the Internet. McLuhan (1964) pointed out that "any technology gradually creates a totally new human environment" (p. viii). Is this true for the Arab world? And if so, what is this new human environment like?

This book will attempt to answer these questions. For the purposes of the study I conducted, an Arab was defined as a citizen of a country that belongs to the Arab League of Nations *and* whose official language is Arabic.[1] The book will cover several major themes, including the current status and infrastructure of telecommunications and information technologies in the Arab world; issues of freedom of expression and how governments of the Arab world are dealing with the online free flow of information on the Internet; Islam and the Internet; usage patterns in the Arab world; and the effects of the Internet on this part of the world.

I will start with an overview of mass media technology and uses in the Arab world to show how the introduction of each new mass medium has contributed some aspects of change to the society. I will then move to focus on the Internet and the opportunities and challenges presented by this new medium.

Broadcast Media Technologies in the Arab World

The Arab world has always reacted favorably to the introduction of new mass media technology. When radio was introduced to this part of the world in the early 1950s, it was a welcomed alternative to newspapers, which served an elite population of literate individuals.[2] In this traditionally oral culture, radio receivers provided much needed functions of information seeking, entertainment, and companionship. Relatively inexpensive radio sets allowed peasants to have the radio on all day while working in the land. Housewives would have their sets on while preparing meals or taking care of children. The availability of this "new" medium at relatively inexpensive cost made it a valuable addition to the Arab home. A few years later, the transistor radio made it even easier to carry the device around, and so radio became even more attractive to the Arab citizen (Boyd, 1999).

The difference between national and international radio broadcasting is not always clear in the Arab world. Arab countries have tried to have powerful medium-wave and short-wave transmitters to be able to reach audiences as far and wide as they could. Arab leaders realized the potential of the new communication medium to their political agendas, as well as to the social and economic development of their countries. Programs intended for domestic audiences are therefore regularly heard in neighboring countries. For example, the national Egyptian radio, one of the most popular and most sophisticated, is heard in most Arab countries, especially at night. Countries with the traditionally strongest radio transmissions, particularly Egypt, and also Syria and Iraq (until the early 1990s), had the strongest propaganda effects on those countries that possessed less sophisticated systems, such as Kuwait, Saudi Arabia, Qatar, Jordan, Lebanon, and the United Arab Emirates (UAE) (Boyd, 1977, 1999).

Patterns of radio use in the Arab world showed different functions for radio based on the type of station or programing that listeners tuned in to. For example, while Arabs tune in to local stations and Egyptian popular stations like *El Shark El Awsat* (The Middle East) and *Sowt El Arab* (Voice of the Arabs) for a mixture of entertainment and information, they have historically tuned in to foreign stations such as the Voice of America (VOA), the British

Broadcasting Corporation (BBC), and the French Radio Monte Carlo Middle East as an alternative source of news to the otherwise national, government-owned and regulated media outlets. Arabs trusted that if there is something their governments are not telling them, they can find out about it from foreign radio stations. This was particularly true before the introduction of satellite television and the Internet, and still is for those who cannot afford such services. Surveys conducted by the BBC show that Arab audiences listen avidly to foreign services, particularly in times of crisis. During the 1991 Gulf War, BBC listening in the United Arab Emirates increased from 21.5% to 51.1%. In Egypt, listening increased from 18.1% to 46.3%. In 2003, Egypt finally allowed private broadcasters into the FM spectrum, a move unprecedented in the Arab world (Media Monitor, 2005).

Similarly, television viewing in the Arab world tends to follow the same patterns as radio listening. One country's national television broadcasting is viewed in neighboring Arab countries. Boyd (1999) reports that "the tall antenna towers on apartment houses and private residences in Jeddah (Saudi Arabia) are not necessary to receive Saudi Arabian television; they are needed to receive Egyptian television from across the Red Sea" (p. 6). Egyptian television is also regularly seen in Lebanon, Syria, the Palestinian territories, and in some Gulf states.

Today, television is an integral part of the lives of Arabs. Most Arab governments realized early on the potential power of television to their development. Even in poorer countries, such as Egypt and the Sudan, television sets were subsidized by the governments to enable the lower socioeconomic classes to acquire them. In the rich Gulf states, large screen projection sets were common by the 1980s. Television ownership is over 90% in all Arab states. More affluent countries such as Saudi Arabia and the United Arab Emirates have ownership rates of more than 99%, with over 50% owning more than one television set (Sakr, 2001). The Gulf states and Egypt have a multichannel television transmission.

Direct broadcast satellite (DBS) technology was introduced to the Arab world in the early 1990s. Satellite adoption skyrocketed within three to five years, particularly in the rich Gulf states, where videocassette recorders were also abundant. The most popular programing on video and DBS has been Egyptian drama, movies, and serials. Arab countries launched their own satellite system, ARABSAT, in 1985. In 1998, Egypt launched its own NILESAT 101, and thus became the only Arab country to individually own a satellite system. This was followed by NILESAT 102 in 2000. Egypt also launched the first

Arab satellite channel, the Egyptian Satellite Channel (ESC), in 1990. Today, all Arab countries have one or more satellite channels. Satellite television was especially welcomed by those who could afford it since it offered an uncensored alternative to otherwise government-owned and regulated media. In Saudi Arabia, for example, there is an official ban on satellite dishes, yet the ban has never been enforced. Satellite dishes and receivers are available for purchase at every local technology store, and numerous satellite dishes appear commonly on top of buildings. In 1997, it was estimated that almost two-thirds of the Saudi population had access to satellite television (International Telecommunication Union [ITU], 2001; Marghalani, Palmgreen, & Boyd, 1998).

Access to satellite television varies from one country to another in the Arab world, depending on the economy of the country and the relative cost of the service. A lack of scientific audience research leaves us at best with estimates of such access. Syria, for example, was estimated in 1999 to have an access of 30% of homes with access to television. Lebanon had an estimate of 42.5% for the same year. Algeria was estimated at 80%. A 1995 survey of students in the United Arab Emirates showed that 81% of the sample had access to satellite television. The average figure for 10 Arab states[3] was estimated at 27.2%. In comparison, Western European states were estimated to have an average of 48.5% (Sakr, 2001). In Egypt, the latest estimates in 2005 put satellite penetration at 28% (Media Monitor, 2005).

While satellite broadcasting offered a way for audiences in the Arab world around government-controlled programing, it has also forced those governments to rethink their broadcasting policies, since it presented competition that they could not impede. For example, when Cable News Network (CNN) first reached the Arab world during the 1991 Gulf War, Egyptian television news broadcasts changed their content and format in an attempt to match those of CNN (Abdulla, 1991). Some argue that the CNN broadcast forced Saudi Arabia to follow up on the news, after it had ignored the Iraqi invasion for three days. Saudis first found out about the invasion from CNN and from the United States Armed Forces Television, which was beaming to U.S. forces stationed in the Gulf (Boyd, 1998; Sakr, 2001). Jordan's ex-Information Minister, Nasser Judeh, reportedly referred to satellite television as "offshore democracy" (Sakr, 2001, p. 4). Audiences in the Arab world finally had access through satellite television to events and/or personalities that the local governments were not willing to feature on their local broadcasting. The power of this new medium was also enhanced by its ability to transcend the boundaries of illiteracy, a demon which the Arab world is still fighting (Sakr, 2001).

Egypt: A Broadcasting Leader

At all stages of mass media development and technology transfer in the Arab world, Egypt has maintained a reputation as a leader, particularly in the development and production of media content. Since the early 1950s, when the country gained independence from the United Kingdom, Egyptian leaders (Gamal Abdel Nasser, Anwar Sadat, and currently, Hosni Mubarak) realized the importance of the media to the nation's development, and to its position as a political and cultural leader of the Arab world. Radio sets, and later television sets, were subsidized by the government, and thousands were made available free of charge for installation in different cultural centers and cooperative units in rural and urban areas. Egypt has been acknowledged as one of the most committed developing countries to the establishment and programing of its mass media.

Ever since radio was introduced to Egypt, it has been integral in serving urban and rural communities alike. However, the impact of Egyptian radio goes far beyond the country's borders. Egyptian radio has always been extremely popular throughout the Arab world. Radio in Egypt started in the 1920s with a few amateur operators. It was not until 1934 when the government, under British occupation, started an official Egyptian radio service, run by the Marconi Company of the United Kingdom. Egyptian radio thrived on acting and musical talent, abundant in the country where a thriving motion picture industry was in existence. Radio programing soon arose to sophisticated levels, and became very popular all over the Arab world (Boyd, 1977).

Shortly after Egypt gained independence from British occupation in 1952, radio, like a lot of other projects and services, was nationalized and Egyptianized. The government realized the potential role of the mass media in the development of Egypt and making it a leader in the Arab world, and so radio soon became the official voice of the government. Programing focused on entertainment and development, and was exported to the rest of the Arab world. Nasser, Egypt's President from 1954 to 1970, realized the importance of radio as a political propaganda device. Through stations like Voice of the Arabs, Nasser broadcast his pan-Arab messages to the rest of the Arab world, which was mostly still under British and French occupation. Nasser enthusiastically supported Voice of the Arabs since its inception in 1953, only one year after Egypt's independence from the British. The station soon became a "household name" (Boyd, 1999, p. 325), particularly reputed for its political news, drama, and even entertainment. Nasser used the station strategically, helping Arab nations gain their independence and serving other "nationalist" causes. Neighboring Arab countries could not respond to Egypt's radio messages, since they did not possess a transmission system as strong and sophisticated as Egypt's. Radio therefore helped Egypt speak to the Arab world, and become a political and cultural leader. Later, after Sadat's 1979 peace agreement with Israel when Arab countries boycotted Egypt, radio was used to counter attacks from these countries (Boyd, 1999; ITU, 2001). Egyptian radio has also maintained an "Overseas Service," broadcasting to foreign countries since Egypt's independence in 1952. Today, Egypt's Overseas Service of Radio Cairo broadcasts to every region of the world in 38 languages.

When television was introduced to Egypt in 1960, the government again was quick to realize its potential power in positing Egypt as a leader, both culturally and politically, in the Arab world. The government issued a presidential decree, which established a series of goals for television and radio broadcasting. Among the goals were elevating the standards of the arts and culture through sophisticated programing, spreading culture among the masses, informing the public about national and international news, informing foreign countries about Egypt and the Arab world, encouraging talent, and providing high-standard entertainment programs. It became government policy to make radio and television sets available to as many people as possible. The Arab culture's emphasis on family and community showed in media consumption patterns. Radio and television set prices were therefore subsidized by the government, and were installed for free in coffee shops, cultural centers, and other public places, especially in the rural parts of the country. Early on, it became a status symbol in rural Egyptian villages to have a television set. The village *omda* (mayor) would acquire a television set even before electricity had reached their region (Abdulla, 2004; Boyd, 1999). In late 1996, television ownership in Egypt was estimated at over 93%, with many owning more than one set (Sakr, 2001). The number of television sets in Egypt in 1997 was estimated by the BBC at 10 million, and by European Audiovisual Observatory at 11.6 million. Television access is available to 98% of urban homes, and more than 90% of rural homes (Boyd, 1999; ITU, 2001).

As a means of realizing its media goals, the government encouraged development communication through the production of cultural and development education programs, which may be one of the few advantages of having a government-owned and operated media system. To fit the traditionally oral culture and the high illiteracy rates, the government particularly encouraged the production of prosocial "drama and programs with a message" that would benefit

Egyptian and Arab audiences. Eventually, development and education through entertainment became part of the culture of radio listening and television viewing in Egypt, and consequently, in the Arab world (Abdulla, 2004; Boyd, 1999; Diase, 1993).

Mass media audiences in Egypt, especially those of the middle and lower socioeconomic classes, turn to television as their number one source of information and entertainment. In our family-oriented culture, audiences also expect to find something educating or some sort of "advice" in most Egyptian serials and soap operas, which have a very wide viewership. Such advice ranges from the handling of social situations and family affairs to more entertainment-education-oriented drama, specifically designed to promote particular prosocial issues. Drama is a main component of programing on Egyptian television (Abdulla, 2004).

Cairo, the capital of Egypt, is known as the "Hollywood of the Middle East." As the leading, most sophisticated radio, television, and motion picture producer in the Arab world, Egyptian programs and dramas are exported to most Arab countries and constitute a main component of overall content on Arab media channels. The Egyptian dialect of spoken, colloquial Arabic is the most widely understood in the Arab world, a direct result of Arabs growing up with Egyptian programs, music, and drama. Even during the Arab boycott of Egypt after the Egyptian–Israeli peace treaty was in effect, Arab countries continued to tune in and import Egyptian radio and television productions (Boyd, 1999; Center for Development Communication, 1989).

The impact of Egyptian media and productions is well demonstrated by, among others, Egyptian singer Um Kalthoum. Known as the Diva of Arabic song and the star of the Orient, Um Kalthoum was reputed to be the power that brought the Arab world together after politics drew it apart. Her influence was especially crucial after the 1967 Arab defeat against Israel. Until her death in 1975, Um Kalthoum had a monthly concert, on the first Thursday of the month. Fans and royalties from all over the Arab world flew in to attend her concert, but more importantly, those who could not afford to come in person would "stick by their radio sets" to listen to the Diva. The Thursday night concert became a social occasion all over the Arab world, where people would gather with their families, friends, or at their local coffee shops to tune in to "the Diva's concert," broadcast live on Radio Cairo. It was known to be the one monthly occasion where almost everybody all over the Arab world shared the same interest and took part in the same activity. Um Kalthoum sang about a variety of topics, from love to Arab nationalism and patriotism. At a time of political trouble and low morale in the Arab world, her songs brought hope and a sense of unity and togetherness to the Arab people.

In the last few years, Egypt has launched several satellite channels, including eight digital Nile Thematic Channels: Nile Drama, Nile Family and Kids, Nile Sports, Nile Entertainment, Nile Educational, Nile Culture, Nile TeleText, and Nile News. Egypt also owns the only Arab satellite channel that broadcasts in English, French, and Hebrew: Nile TV International. Launched in 1993, Nile TV International targets foreign viewers in Europe, North America, Asia, and Africa. The channel aims at promoting Arab culture and literature, as well as expressing the Arab political view points. In an important development, Egypt has also permitted three private satellite channels so far, Dream I, Dream II, and *El Mehwar*, to start broadcasting to Egyptian and Arab audiences. Egypt is currently the number one country in the Arab world in terms of the number of satellite channels and hours of broadcasting. Egypt's own satellite, NILESAT 101, was launched in 1998, with over 80 television and 400 radio channels. The success of the project led to the launch of another satellite, NILESAT 102, in August 2000.

Egypt has also completed work on the Media Production City (MPC), located in the Sixth of October city, just southwest of Cairo. MPC is built on an area of 35 million square meters (more than 600 feddans), and is considered the Arab world's equivalent of Hollywood. The city, which features 114 state-of-the-art studio complexes and numerous temporary and permanent location sets, aims at achieving an annual 3,500 hours of indoor studio production, and 5,000 hours of programs and outdoor production. It aims at producing 100 long motion pictures per year. MPC is also equipped with a 2,500-seat hall for the television production of dramatic performances and concerts, several open-air theatres, a film-processing laboratory complex, a 4-star, 250-room hotel, and a training center with the capacity to train 3,000 individuals per year in all aspects of television and film production. Among the built-in locations of the city is a simulation of *Tal El Amarna*, the Pharoanic capital of old Egypt; replicas of several important historical landmarks including the Pyramids and the Sphinx, Saladin's Citadel, numerous Pharoanic temples, numerous Islamic and Coptic monuments, and Bibliotheca Alexandrina; an Islamic city in the Fatimid and Mamluk style; and an underwater film-shooting area. A children's village, Magic Land, provides an amphitheatre where dolphin and water ballet shows are performed daily. The studio complexes were fully occupied months before the city was officially inaugurated in 2001 (Media Production City, n.d.).

· 2 ·

TELECOMMUNICATIONS AND INFORMATION TECHNOLOGIES IN EGYPT AND THE ARAB WORLD

The decade of the 1990s was an important one for mass media and communication technologies in the Arab world. In addition to the boom in television, satellite and VCR technologies, other technologies, namely telephony and computers, became better integrated in this part of the world. The numbers, however, remain relatively low compared to other parts of the developed world. As the internet is mainly composed of connections of telephones and computers, it becomes important to take a closer look at the structure of information and telecommunications technology in the Arab world.

The average number of telephone lines in the Arab world has more than doubled in the 1990s. As shown in Table 2.1, the number was estimated at 94 telephone mainlines per 1,000 people in 2003, a big increase from 35 per 1,000 people in 1990. Cellular phones were estimated at 118 per 1,000 in 2003, an even bigger increase from 38 per 1,000 just three years earlier in 2000 (United Nations Development Programme [UNDP], 2005). This gives an indication of how widely new technologies are spreading throughout the Arab world, particularly in the more affluent countries.

Wide variations in telecommunication capabilities exist within the Arab world. The United Arab Emirates leads the way with 281 telephone mainlines and 736 cell phones per 1,000 people in 2003. Countries such as Kuwait,

Table 2.1. Main land and cellular telephone lines in the Arab world and selected world countries in 1990, 2000, and 2003

	Telephone mainlines per 1,000 people			Cellular phones per 1,000 people		
Country	1990	2000	2003	1990	2000	2003
Algeria	32	57	69	<0.5	3	45
Bahrain	191	249	268	10	430	638
Egypt	30	86	138*	<0.5	21	195*
Iraq	39	30		0		
Jordan	72	92	114	<0.5	58	242
Kuwait	188	277	196	12	249	572
Lebanon	155	195	200	0	212	234
Libyan Arab Jamahiriya	48	108	136	0	7	23
Morocco	16	50	40	<0.5	83	244
Oman	60	89	88	2	65	228
Occupied Palestinian Territories			87	0		133
Qatar	220	268	261	9	202	533
Saudi Arabia	77	137	155	1	64	321
Syrian Arab Republic	41	103	123	0	2	68
Sudan	3	12	27	0	1	20
Tunisia	37	90	118	<0.5	6	197
United Arab Emirates	224	293	281	19	548	736
Yemen	11	19	28	0	2	35
Arab States	**35**	**77**	**94**	**<0.5**	**38**	**118**
Compare with						
United States	547	700	624	21	398	546
Norway	502	731	713	46	751	909
France	495	579	566	5	493	696
Sweden	681	682	736	54	717	980
United Kingdom	441	589	591	19	727	912
Japan	441	544	472	7	569	679
World	**81**	**163**	**184**	**2**	**121**	**226**

Note. Data compiled from MCIT, 2006; UNDP, 2002, 2005; UNDP & Arab Fund for Economic & Social Development (AFESD), 2002, 2003, 2004.
* Information as of June 2006.

Lebanon, Bahrain, and Qatar maintain relatively high numbers, while other countries such as Sudan, Yemen, Iraq, and the occupied Palestinian territories have very low numbers. Egypt enjoys a high rate of growth in telecommunications technology. In 2003, the number of telephone mainlines was estimated at 127 per 1,000 people compared to 86 just three years earlier in 2000. Egypt's mobile phone lines had an even higher growth rate. The numbers increased from 21 mobile phones per 1,000 people in 2000 to 84 mobile phones per 1,000 people in 2003 (UNDP, 2005). Today, Egypt has 138 mainlines and 195 mobile phones per 1,000 people (Ministry of Communications and Information Technology [MCIT], 2006).

The high growth rates in cellular telephony in the Arab world are worth taking a closer look at. The figures are staggering in some countries, where the number of cell phones per 1,000 people is more than double the number of regular mainlines. As mentioned before, this is the case in the United Arab Emirates, and also in Bahrain, which has 638 cell phones per 1,000 people, compared to 268 mainlines. Kuwait also has 572 cell phones per 1,000 people, and 196 mainlines. Next on the list is Qatar, with 533 cell phones per 1,000 people and only 261 mainlines. Jordan again has 242 cell phones per 1,000 people, and 114 mainlines. The numbers are particularly interesting in Morocco, with 244 cell phones per 1,000 people, and only 40 mainlines. In most of these countries, the average number of telephone mainlines is slightly decreasing from year to year, while the number of cell phones is increasing, an indication of people converting to new telecommunication technologies. Overall, the number of cell phones per 1,000 people in the Arab world is higher, estimated at 118, than regular mainlines, estimated at 94. This phenomenon is comparable to overall world figures, which stand at 226 cell phones per 1,000 people, and 184 mainlines. The difference, of course, is in the quantity of lines, rather than the ratio of cell phones to mainlines. It is worth mentioning though that only the highly advanced world countries, such as Norway, Sweden, the United Kingdom, and Japan, have more cell phones per 1,000 people than regular mainlines. In the United States, mainlines are still ahead, with 624 mainlines per 1,000 people, compared to 546 cell phones.

The World Economic Forum (2004) reported that the region of the Middle East and North Africa (including Iran and Israel) has witnessed a 340% increase in cellular mobile telephony use between 1999 and 2002, which amounted to 25.8 million new subscribers. Egypt has the highest number of main telephone lines among Arab states, and the third highest number of cellular phone subscribers. Egypt led the way between 1999 and 2002 with 2.7 million new main telephone lines, and 4.1 million mobile phone subscribers. At the end of 2005, there were 10.4 million telephone mainlines and 14 million cell phone subscribers in Egypt (MCIT, 2006).

In terms of growth, Jordan had an amazing growth rate of 930% between 1999 and 2002, bringing the total number of subscribers to 1.2 million in 2002. Tunisia had a growth rate of 812% between 2000 and 2003, from 6 mobile phone lines per 1,000 people in 2000 to 52 in 2002. The total number of mobile subscribers in 2002 was 503,911.

Algeria had a growth rate of 456% in mobile subscribers, from 3 per 1,000 people in 1999 to 13 per 1,000 people in 2002. The number of subscribers, however, was still 400,000. Yemen and Syria both had excellent growth rates, bringing the number of cell phone users per 1,000 people from 2 in 2000 to 21 and 23 respectively in 2002.

Privatization of Telecommunications Sectors in the Arab World

Since the late 1990s, there has been a trend all over the world to privatize and liberalize telecommunications sectors. The Arab world has been no exception. For the most part, privatizing efforts have been clearer in mobile telephony services than in regular mainlines. Still, several Arab countries have managed to privatize both systems of mobile and fixed mainlines. These countries include Egypt, the United Arab Emirates, Bahrain, Jordan, Morocco, Qatar, and Yemen.

The United Arab Emirates has gone ahead with big strides in terms of privatization and also in terms of building its telecommunications infrastructure. The main player, Emirates Telecommunications Corporation (*Etisalat*), is 40% owned by shareholders. The company has been around since 1976, serving the UAE population of fewer than 4 million. *Etisalat* is one of the major telecommunications companies in the Arab world, with a Global System for Mobile Communications (GSM) service that started in 1994. Today, it serves 3 million subscribers, at a more than 75% penetration rate. *Thuraya* Satellite Telecommunication Company, in which *Etisalat* is a major shareholder, is the key private player in charge of satellite-based mobile telephony (Etisalat, 2006). *Thuraya* was founded in 1997 by a group of leading national telecommunications operators and international investors. They have since expanded to carry out projects in other Arab countries such as Morocco, and Oman, as well as in Afghanistan, Kenya, and Tanzania

(Thuraya, 2006). Telecommunications services in the United Arab Emirates are regulated by the Ministry of Communications.

In Bahrain, which also enjoys an advanced infrastructure, Batelco (Bahrain Telecommunications Company) is the key player. The company was established in 1981 as a Bahraini shareholding company. The Government of Bahrain owns 39% of the shares, while 20% of the shares are owned by the British firm Cable and Wireless (C&W). The rest are owned by other financial and commercial organizations, and traded on the Bahraini Stock Exchange for the benefit of Bahraini and Gulf Cooperation Council citizens (Batelco, 2006).

In Jordan, the Jordan Telecommunications Corporation (JTC) was established in 1997 as a government-owned company to operate fixed mainline telephony in the country. However, in 2000, 40% of the company's shares were sold to France Telecom and Arab Bank. The company had started to provide mobile telephony services in 1998, competing with Fastlink, which is owned by the Egyptian Orascom Telecom. Telecommunications in Jordan is regulated by a special body, the Telecommunications Regulatory Commission (Dahel, 2001; JTC, 2006).

Morocco offers a good example of deregulation and privatization in the Arab world. It took the country seven years of planning for the Post Office and Telecommunications Act to finally be approved by the parliament in 1997. The act aimed at providing major reforms to the telecommunications sector in the country. In doing so, its main objectives were to

- provide a transparent and effective regulatory framework that promotes fair competition,
- pursue the development of telecommunications networks and services by adapting them to the new developments in technology,
- provide the economy with the needed telecommunications services in order to promote and open up the economy to international investors and help its integration into the world economy,
- promote the creation of jobs in the telecommunications sector to serve the public and the economy.

A new regulatory body was also created as a result of the Post Office and Telecommunications Act. The National Telecommunications Regulatory Agency (ANRT) was therefore responsible for monitoring market players, making sure the policies are implemented, as well as resolving any potential disputes. The key player for telephone operations became Maroc Telecom (*Itisalat Al Magrib*). But more importantly, the highly organized and liberalized structure, backed by the Monarchy, brought about a sleuth of foreign investments. Vivendi Universal bought 35% of Maroc Telecom shares for US$2.3 billion. Shortly after, Medi Telecom Consortium paid US$1.1 billion to become the second mobile telephony operator in the country. The mobile telephony market hugely expanded in terms of subscribers as well (ITU, 2001).

Yemen has a joint venture company between the Yemeni Public Telecommunication Corporation (49%) and the British Cable and Wireless (C&W) (51%). The resulting Teleyemen is the major player in the telecommunication market in the country, operating both fixed mainlines and mobile telephony under the auspices of the Ministry of Communications. In January 2004, France Telecom took over the management of Teleyemen for five years. Another private company, Sabafon, started operating in the mobile telephony market in 2001 (Dahel, 2001).

In Qatar, Qatar Public Telecommunications Corporation (Q-Tel) is the exclusive telecommunications operator, in charge of both mainline and mobile telephony systems, in addition to cable television and Internet services. The company offers 45% of its shares for trading on the Doha Securities Market. It is also listed on the London Stock Exchange, and the Bahrain and Abu Dhabi (United Arab Emirates) stock markets (Dahel, 2001; Q-Tel, 2006).

Another Arab country that offers a good example of telecommunications privatization despite its poor economic resources is the Sudan. The country believes that economic development is directly linked to the rate of telephone penetration. This belief underlined the privatization efforts of the telecommunications sector by the government. In 1993, the Sudan Telecommunications Company (Sudatel) was established as a public shareholder company, between the Sudanese Government and 37 founder shareholders from the private sector. The initial capital of US$150 million was soon raised to US$250 million. Sudatel shareholders include the United Arab Emirates' *Etisalat*, Q-Tel of Qatar, as well as other GCC investors (Sudatel, 2006).

Sudatel aimed at accelerating the telecommunications market development. The objective was to raise telephone penetration from 3 lines per 1,000 people in 1990 to 10 lines in 2000. The company surpassed its objectives, with a penetration of 12 lines per 1,000 people in 2000, and 27 lines per 1,000 people in 2003. Sudatel is currently listed on the Khartoum Stock Exchange as well as the Bahraini Stock Exchange (Sudatel, 2006; UNDP, 2005).

Some Arab countries have private mobile telephony operators while their fixed mainlines are still operated by a government-owned entity. These countries include Algeria, Kuwait, Lebanon, and Syria. In Syria, the Syrian Telecommunications Establishment (STE) operates fixed mainline telephony. Two private companies, SyriaTel, owned by the Egyptian Orascom Telecom, and Investcom, a Lebanese company, operate the mobile telephony market. Both companies have a contract through which they have to transfer 30–50% of their annual profits to the Syrian government (Dahel, 2001).

In Lebanon, the Ministry of Telecommunications adopted a new policy and law in 2004 that aims to enhance liberalization and privatization efforts and establish a new regulatory body for the sector. The policy also aims to optimize the telecommunications service in the country and the value of the state-owned operators. The ultimate goal is to enhance the information society in Lebanon, and to attract foreign investments and create additional jobs. Currently, the government entity Liban Telecom is the mainline operator. Mobile telephony is currently operated under an interim agreement with FTML and LibanCell. The Lebanese government recently announced an international public tender for the two mobile businesses, and is entertaining the idea of introducing a third company to the market, a subsidiary of the government-owned Liban Telecom (Lebanese Ministry of Telecommunications, 2006).

The situation in Kuwait is quite similar to that of Lebanon, with a main governmental player, in this case the Kuwaiti Ministry of Communications, operating the mainline telephony services. Two companies operate the mobile telephony arena: Mobile Telecommunications Company (MTC) and the National Mobile Telecommunications Company (NMTC) (Dahel, 2001).

In Algeria, the Ministry of Posts and Telecommunications operated both mainline and mobile telephony until 2000 when a law was passed to reform and liberalize the sector. Algeria Telecom was established to become the sole operator of mainline telephony and a co-operator of mobile telephony. A second mobile operating license was obtained by Orascom Telecom, and started operating in Algeria in 2002 (Dahel, 2001).

A few Arab countries are more cautiously moving toward privatizing their mobile operators, while keeping their mainline telephony under government operation. These include Oman, Tunisia, and Saudi Arabia. Oman Telecommunications Company (OmanTel) has been announcing the expected public offering of a percentage of its shares to Omani citizens for more than two years now. It is expected that the shares will be on the Omani capital market in June 2006 (*Hatef Telecom*, 2006).

Saudi Telecommunications Company (STC) or Saudi Telecom (*Al Etasalat Al Saudia*) was established in 1998 as an extension of the Saudi Ministry of Post,

Telephones, and Telegraphs. In 2002, 30% of its shares were offered to Saudi citizens, thus taking its first steps toward privatization (Saudi Telecom, 2006).

Telecommunications and Information Technology in Egypt

Since 1854, telecommunications in Egypt was dominated by one key player, The Arab Republic of Egypt National Telecommunication Organization (ARENTO), a government-owned and operated entity. In April 1998, as part of a privatization plan, ARENTO was taken over by Telecom Egypt, a joint-stock company in which the Egyptian government owns the majority shares. Telecom Egypt is the biggest telecommunications operation in the Arab world, with 10.6 million subscribers in June 2006, and an annual growth rate of 17% since 1996. The penetration rate for telephone mainlines is 14.8%. Telecom Egypt has a monopoly over landline telephone services. There are also three private payphone operators in Egypt: Menatel, Nile Telecom, and Telecom Egypt. In June 2006, the companies had a total of 55,494 pay phones (ITU, 2001; MCIT, 2006; UNDP, 2005).

In 1997, Egypt introduced mobile telephony. Less than a year later, the country's GSM network was acquired by MobiNil Egypt, a private company owned by the Orascom Telecom consortium. MobiNil is the leading mobile telephony provider in Egypt, with more than 7.5 million subscribers in June 2006. MobiNil service covers more than 91% of Egypt, with roaming service available in 112 countries (MCIT, 2006; MobiNil, 2006). Another private company that offers mobile telephone services in Egypt is Misrfone Telecommunications. Operating first under the brand name ClickGSM and then Vodafone Egypt, the company launched its service in late 1998. The Vodafone network currently covers 98% of Egypt, with roaming services in 117 countries. The company has over 7.1 million subscribers (MCIT, 2006; Vodafone Egypt, 2006). The total number of mobile phone subscribers in Egypt was up to about 15 million in June 2006, a penetration rate of almost 20%.

In February 2006, the Ministry of Communications and Information Technology (MCIT) and the National Telecommunications Regulatory Authority (NTRA) announced accepting proposals for a third mobile telephony company. On the first day of its availability, ten multinational telecommunication consortia purchased the US$25,000 specifications handbook for the company. In July 2006, the contract was awarded to a consortium led by the

United Arab Emirates' *Etisalat*. The winning bid was in the amount of US$2.9 billion, a huge contrast to the US$500 million paid by each of the other two companies in 1998. Egyptian government officials brag that the difference shows the strength and growth of the Egyptian telecommunications industry. *Etisalat* leads the consortium by 66% of the shares, while the rest is owned by the National Post Authority (20%), the National Bank of Egypt (10%), and the Commercial International Bank (4%). The consortium will also pay 6% of its revenues to the NTRA. MCIT expects that competition from the third mobile telephony company will lead to improved services and a bigger market, driving the subscriber base to 35 million in five years.

Since Egypt's independence in 1952, the government body responsible for telecommunications in the country was the Ministry of Transport and Telecommunications. However, in 1998, the government created a new MCIT. The new ministry was headed by Dr. Ahmed Nazif, who in July 2004 became prime minister of Egypt. Nazif is a former CEO of a large private corporation, and a former professor of computer engineering, with a PhD from McGill University in Canada. With such diverse experience, Nazif managed to steer the ministry in the direction of deregulating and privatizing mobile telephony and Internet service in Egypt, and provided much-needed new telephone services in the country, such as call-collect, 0-800 (toll free numbers, telephony fees paid by the receiver), and 0-900 numbers (pay per call, higher fees per minute for the caller). The quality of telephone service was greatly enhanced by switching all lines from analog to digital. Customers also benefited from a reduction in rates for international phone calls by 50% in 1999, and by a further 30% in 2000. The introduction of Voice over Internet Protocol (VoIP) led to another reduction in prices, so that now the price of a phone call to the United States from Egypt during off-peak times is only 90 pts (US$0.15). According to the United States Information Service (USIS) (2000), "the communications and information infrastructure has improved significantly in Egypt over the last few years."

When Nazif became Egypt's prime minister in July 2004, Tarek Kamel took over the MCIT. Kamel was instrumental in bringing the Internet to Egypt in 1993, and had been working closely with Nazif as a consultant for years. As such, the MCIT policies did not suffer any changes as a result of the cabinet reshuffle. Kamel had been the head of the Internet Society of Egypt, and is the only Arab member in the Internet Corporation for Assigned Names and Numbers (ICANN). He is also credited to be the main figure behind the free Internet initiative that Egypt introduced in January 2002.

Computer ownership is also on the rise in Egypt, despite the relative high cost of computers. Reports by the USIS indicate that there are more than 200,000 new personal computers purchased every year in Egypt. Three-quarters of that number are assembled in the country (USIS, 2000). The Egyptian government now has an ambitious plan to make computers available in every Egyptian home, especially for students and academicians, for a monthly payment plan of about as little as 60 L.E. (US$10) a month. In the two years following the start of this service in November 2002, over 80,000 computers have been sold (MCIT, 2006). The overall computer market in Egypt is estimated to be growing at 50–60% per year (Alterman, 1998).

There are about 1,600 information technology (IT) and communications companies in Egypt, some of which specialize in producing Arabic-version products for the Egyptian and Arab markets. These companies employ a highly technical information technology workforce in the private sector, estimated at more than 42,000 people (MCIT, 2006).

Numerous international computer and information technology companies have local and regional offices in Egypt. These companies make use of the Egyptian IT workforce. International companies with offices in Egypt include PSINet (network switching facility), Intel, Network Solutions (Web addressing), Hewlett Packard, Sun, Lucent Technologies, IBM, Cisco, Siemens, Microsoft, and Oracle (USIS, 2000). These companies are now all part of Egypt's new Smart Village, one of the latest efforts to boost the IT industry in Egypt. The village, on the outskirts of Cairo, is designed to be "a technology park with state-of-the-art infrastructure geared towards hi-tech businesses" (Smart Village, 2006). Backed by the support of the Egyptian government and major IT businesses, the village encourages conglomerates of IT and Internet businesses through a wide variety of technological and administrative services, an impressive infrastructure as well as financial investment and tax benefits.

Egypt has been trying to enhance its IT workforce by providing facilities and training opportunities in schools and on the job. In 2000, Egypt signed an agreement with Microsoft to allow university students to use Microsoft software packages for a nominal annual fee of L.E. 30 (US$6.5). In the same year, Lucent Technologies opened a new regional office in Cairo, housing 200 local computer engineers. Lucent Technologies planned a US$10 million budget to train its new workforce in the fields of software development and wireless communications. Another major IT player that expanded its operations in Egypt was IBM, who, in 2002, increased the local workforce for its regional offices in Egypt by 25% (US-Egypt President's Council, 2002; USIS, 2000).

In September 2002, MCIT, Telecom Egypt, and Germany's Siemens signed a five-year agreement, aiming at modernizing and developing the information and communication industry in Egypt. According to the agreement, Siemens is to invest €72 million (US$77.8 million) to establish training centers in Egypt to support the company's regional activities, and to establish an "IP Center of Competence," which will undertake the development of multimedia applications in cooperation with Egyptian computer engineers and businesses. The agreement is expected to increase Egypt's export of telecommunications products and services to about €120 million (US$129.7 million) over five years, thus serving to make Egypt an information technology center for the Arab world (US-Egypt President's Council, 2002).

The MCIT has an ambitious plan to further enhance its IT human resources, with an allocated annual budget of L.E. 195 million (US$42.4 million). So far, the different ongoing training programs have produced some 143,000 graduates who trained on basic IT skills. The MCIT also signed agreements with 116 colleges and universities in Egypt to train some 6,000 IT students per year (MCIT, 2006; US-Egypt President's Council, 2002). Egypt's efforts to enhance IT in education have been described as "unmatched among the developing countries in the region" (Palmgren, 2001).

The Effects of Communication Technology on Egypt and the Arab World

The introduction of each new medium has carried some effects on the fabric of the Arab world. For example, as shown in chapter 1, Nasser used radio, and later television, to spread his pan-Arabism socialist messages, and to position Egypt as a political and cultural leader in the Arab world. With the introduction of satellite television into the Arab world, citizens now have a much broader access to information. A survey of satellite dish owners showed that they bought their receivers to gain access to the outside world, to learn about other cultures, and to have access to news and information from outside sources. Their news preferences came from Arab independent news channels such as MBC (Middle East Broadcasting Corporation) and Al-Jazeera (Shaheen, 2001), as well as foreign channels such as CNN, Euronews, and the BBC (Youssef, 2001). On a social level, some studies (Elewa, 1998; Qulaini,

1997) showed that watching satellite television is linked to a desire for materialism and Westernization among Egyptian youth, including a desire to leave the country, to indulge in fast food, and to follow the latest in Western style and fashion. As for the Internet, while it might be early to assess the effects of this new medium on the Arab world, some scholars believe the potential for transformation has never been greater.

Masmoudi (1997) argued that the Internet would have serious positive effects on the Arab world on all fronts. Politically, he anticipated that the free flow of information that is bound to be enhanced by access to the Internet would force Arab leaders to be more open and allow for more freedom of speech in their countries. He argued that such free flow of information would inevitably affect public opinion, and eventually, would have an effect on political decision making. Thus, access to the Internet would eventually lead to a more democratic, more decentralized, and more open political system.

Alterman (1998) reiterated the same point. He argued that with a multitude of information sources available to them, Arabs would have to be selective and evaluate the credibility of their sources. Governments would then be compelled to compete with those other sources, and therefore would have to present information whose credibility is acceptable to the public.

On the economic track, Masmoudi (1997) also argued that the IT needs brought about by the Internet would encourage governments in the Arab world to take on more private sector projects. A new generation of young, well-trained IT personnel would emerge to play a major role in the economy of these countries. As discussed above, these efforts have already begun, particularly in countries such as Egypt and the United Arab Emirates.

In terms of education, the Internet offers an infinite wealth of information on the Web, and a whole new world of distance learning and information exchange. Competition would force Arab governments to reexamine their own educational policies, and update their curricula to include more information as well as specialized courses on computers, IT, and the Internet.

As for cultural exchange, Masmoudi (1997) argued that if Arabs catch on to the information revolution on the Internet, they should be able to learn more about the outside world, as well as teach the outside world about Arab culture. In this way, he argued that the Internet would act as a gap-leveler, bridging the gap between the West and the Arab world, and enabling Arabs to become a more active part of McLuhan's (1962, 1964) global village.

Abdel Raouf (1998) argued for the importance of creating an information society in Egypt, and stressed that social factors define the role of information in any given society. He further argued that the effects of the Internet on the Arab culture would be defined by how Arabs choose to use the new technology, and how fast they adopt it. Abdel Raouf also argued that the IT industry brings about the potential of advancements in management, the labor force, and hence the economy in general. Like Masmoudi (1997), Abdel Raouf also pointed out the potential of the Internet in mobilizing the Arab civic society and creating a more democratic atmosphere in which the average individual has a bigger say in the political arena.

Alterman (1998) pointed out to the potential role of the Internet in promoting intra-Arab trade and technology transfer. He argued that since the Internet eliminates physical distances between businesses, particularly if the product is online technology, Arabs should be able to capitalize on their IT and Internet experts, and conduct intra-Arab business online. Potential products include Web design, Web hosting, and Web page maintenance.

Nagib (2000) argued that using Arabic characters on the Internet was likely to "draw Arabs together and unite them within a common framework of ideas and interests." She projected that as Arabs and Asians become more productive in uploading information onto the Web, English would cease to be the dominant language on the Internet. She argued that "English was predominant over the Internet, (and so) it had every right to gain the attention of the world, but this does not necessarily have to be the case anymore" (pp. 144–145).

While the Internet might be likely to make the world more of a global village, some Arabs still approach it with caution as they try to assess whether this will mean a one-culture village where Westernization is the order of the day, or a multicultural village where more understanding and tolerance of their identity and causes could be promoted. For example, Amin (2000) reported that "Open access to the world of the Internet is seen by many as a threat to Arab culture, values, and traditions." On the other hand, Masmoudi (1997) said the Internet has the potential to bridge the cultural gap between the Arab world and the West. Nagib (2000) argued that the Internet would provide Arabs with a chance to "expose themselves before an international audience" and "diffuse their culture and heritage worldwide" (p. 146).

Perhaps one of the more critical aspects of the government-owned media systems in Egypt and the Arab world is their lack of participation by the masses. While they have proved to be an effective channel for governments and leaders to reach the public and influence the masses, the media have not engaged enough active public participation. The *Egypt Human Development Report 2003* and *2005* criticized the media for their lack of stimulation in terms of organizing civil society and influencing public opinion and/or decision makers. The report called on the media in Egypt and the Arab world to play a more active role in reinforcing and reshaping culture and in increasing political and civil participation (UNDP & Egypt's Institute of National Planning, 2003, 2005). In this regard, the Internet, with its obvious advantages of interactivity and easy access, can play an important role in providing a voice for the public. Egypt's efforts so far as an established mass media leader in the Arab world is an indicator of the potential role it can play in providing Arab-related content and services both in English and in Arabic on the Internet, as well as in serving as an IT center, and eventually, as a model for political and civil participation in the Middle East and the Arab world.

· 3 ·

A BRIEF HISTORY OF THE INTERNET

The Internet is a network of networks. It is a huge collection of computer networks that are connected together via telephone lines through central nodes and can thus communicate with each other. The network can therefore carry text, picture, or audio and video messages, and transfer them from one place to the other, even if the two places are a few meters or a few continents apart, just as long as they are both connected to this huge network, the Internet.

The history of the Internet goes back as far as the early 1960s when the United States Department of Defense decided to sponsor military research about new modes of communication under severe military conditions. This was a national security need brought about by the challenges of the time. During that time, the Cold War was in full gear, and the Soviet Union had just sent the Sputnik satellites into space. The United States felt threatened by the Soviet supremacy in this regard, and decided that it had to achieve some technological advances in space technology, but it also paid special attention to coming up with a system of communication that would allow the important departments of the U.S. government to communicate in case of a Soviet nuclear attack.

The known modes of communication at the time were designed following what is known as a circuit-switching model, whereby a message was transferred across central points to a destination. The problem with this model was that if a central point was hit during a nuclear attack, there would be no way of getting the message across. To illustrate, suppose we need to get a message across from Egypt to Morocco, and suppose the only way to do this was by having someone carry it and travel with it across North Africa on a particular route: Egypt to Libya, Algeria, and then Morocco. If there is a problem with the travel path from Libya to Algeria, for example, the message would not reach its final destination. The message would have been blocked on the way.

The United States realized that an alternative system had to therefore be designed to account for a danger of that kind. And thus telecommunications scientists thought of a new system that was later to become the initial idea of the Internet. The alternative system was that of packet switching, pioneered separately by Donald Davies in Great Britain and Paul Baran in the United States. It was mainly based on a system where the message is transferred in packets across alternative points of routing. Packet switching breaks a message down into packages, then attempts to send each packet of the message separately. The packets are disassembled from the message, and sent to their final destination in no particular order using multiple alternative routes. The main advantage to this strategy is that if a communication node was hit in a military attack, the system would simply route the packet to a different node on its way to the final destination. The message, therefore, is not lost. Again to illustrate by using the same example we used before, this time we will break down the message, so that each paragraph of it is sent separately with a person across any random route of that person's choice. If now one person finds the road blocked between Libya and Algeria, that messenger can change paths and take an alternative route, say through Tunisia in the North or Niger in the South. Once all the packets reach their final destination, they are then rearranged to appear in the same order and format in which they were originally sent.

Packet switching has other advantages. For one thing, it makes it almost impossible to spy on a particular message, since all messages are broken down and the parts are sent in no particular order. The technology also makes for more efficient time-sharing on a network. With circuit switching, a message has to wait for the line to clear before it gets sent. So if you were sending a short message, and there were several long messages in line to be sent, it would have to wait for its turn. With packet switching, this is not the case. Since packets are sent in no particular order, and there are plenty of alternative routes to send them, and since a short message constitutes fewer packets than a long message, the fewer packets have a chance of getting to their destination much faster.

In 1968, packet switching technology was embraced by ARPA (the Advanced Research Projects Agency), a unit formed by the U.S. Department of Defense, onto which funds were pouring under directives from the then U.S. President Lyndon B. Johnson, to help advance knowledge and IT in the United States and face the competition of the Soviet empire. To reach this goal, ARPA attempted to connect a network of scientists and researchers across different universities in the United States. Early examples of connected computers were available at research centers at the University of California at Los Angeles, the University of California at Santa Barbara, Stanford Research Institute, and the University of Utah. It was therefore decided in 1969 that these four centers would be the first experimental link into what was known as the ARPANET, which was the first form of the communication network that grew to be the Internet.

By 1971, e-mail was formally introduced when the first special software was developed to send messages across the system. A year later, ARPANET gave its first public demonstration of the network with 40 computers connected to the system. And by 1973, the Internet finally became an international system when University College of London, England, and Royal Radar Establishment of Norway joined the system. Queen Elizabeth II sent the first celebrity e-mail in 1976.

Two researchers, Vinton Cerf and Robert Kahn, were instrumental during the following two decades in developing the technical aspects of the network and coming up with technology protocols to standardize it across different systems and make it more efficient. Their work produced what became known as Transmission Control Protocol (TCP)/Internet Protocol (IP), which is the leading technological aspect of the way the Internet works today. Simply, each computer now has what is called an IP address, which is a series of numbers that would work to clarify where the message is sent to, just like a postal address would clarify to the postman where to deliver a letter. IP is responsible for assigning a final destination to each of the packets of a message. TCP is responsible for breaking down the message, as well as the actual transfer of the packets across different nodes along the way. At the receiver end, TCP is responsible for putting the packets back together to form the original message. Then IP makes sure it is delivered to the final destination. So if you were to send me an e-mail, it would be broken down by TCP into packets at your end, and stamped with my address by IP. Then TCP would transfer the packets individually across the network. At my end, TCP would make sure the message is put back together as you originally sent it. Once all the packages of the message are received, IP would make sure it gets in my mailbox. TCP is also responsible for higher order services, such as the retransmission of a packet or part of a message if it gets lost or destructed along the way.

Eventually, with the number of users increasing on the Internet and with the introduction of personal computers (which meant user numbers would multiply even more), the U.S. military was concerned about the possibility of hackers breaking into the system. Therefore, the decision was made to separate the military part of the network, called MILNET, from the academic part, which remained, to be called ARPANET. The military then gave control of ARPANET in 1982 to the U.S. National Science Foundation (NSF).

The year 1984 was another important landmark for the Internet. To begin with, the number of hosts reached 1,000. And a system known as Domain Name Server (DNS) was introduced. DNS is the system that determines how Web addresses are assigned. The domains are assigned depending on the type of Web site. There were six main types of domain names, each with identifying letters: .org for organizations, .net for networks and nongovernmental agencies, .com for commercial sites, .mil for military, .gov for governmental sites, and .edu for educational sites. Different countries were also given a two-letter prefix for identification purposes, for example, .lb for Lebanon, .sa for Saudi Arabia, and .eg for Egypt. The only exception was the United States, who did not use a prefix, since the Internet was mainly an American invention (the same way that English stamps do not specify a country name since Britain had the first postal service system).

And so the Internet developed over the years to what its original founders probably never conceived of. Although it started with a vision to connect computers for purposes of hardware, software, and data sharing, the users of the network took matters into their own hands in deciding what the most popular and appropriate uses for it would be. What was unique in this case was that the users of the network were often its developers. This allowed for maximum flexibility and served to better fulfill the needs of those users. Connections continued to be made to sites overseas where separate smaller networks were in use particularly in Japan, France, the United Kingdom, Switzerland, and some other European countries. Other networks such as BITnet (Because It's Time Network) and EARN (European Academic and Research Network), which were in use since the 1980s, were integrated into the system. TCP/IP became the standard protocol for the network of networks. The ARPANET backbone was getting old, so it was replaced by the NSFNET backbone and eventually become known as the Internet.

Eventually, the Internet was privatized in the early 1990s, and Internet service providers (ISPs) were allowed to offer access to paying customers through the ISP's own backbones. Factors that led to the exploding of the Internet included the introduction of personal computers, and TCP/IP, but also people like Tim Berners-Lee of CERN (European Organization for Nuclear Research), who in the early 1990s introduced the idea of the World Wide Web (WWW) as the first interface utilizing hypertext. Berners-Lee, a graduate of Oxford University in the United Kingdom, invented HTML (HyperText Markup Language), the language used to build Web pages on the Internet. Before the Web, information on the Internet was available on Gopher, a text-based menu-driven interface, which was introduced in 1991. Although Gopher was in itself a big step, it was still very difficult to reach a particular topic of interest on the Internet unless the user knew exactly where it was. The introduction of the Web and HTML led to an information explosion on the Internet since it made it so much easier for users to learn the language and then post information on the Web. Berners-Lee is now director of the World Wide Web Consortium (W3C), a forum of organizations that aim to realize the full potential of the Web to its users.

Another major contribution to the Internet came in 1993, when Marc Andreessen introduced Mosaic as the first color graphic interface browser. Andreessen was a student at the University of Illinois's National Center for Supercomputing Applications (NCSA), and became very interested in Web technology. Besides the obvious advantage of introducing color, Mosaic facilitated dealing with pictures on the Web, and made it possible for a Web user to go to another Web page by simply clicking on the link to that page. The software was available for download on the Internet and became so extremely popular that the *New York Times* featured it on the front page of its Business section, crediting it with a potential revolution in the world of Internet use. Mosaic, however, was owned by NCSA since the software was developed at the university, and so upon graduation, Andreessen was motivated to develop an even better software that he could claim as his own. In 1995, he and his team developed Netscape, which was acquired by AOL (America Online) three years later. Microsoft rose up to the competition and developed Internet Explorer. These two leading browsers were a real revolution in facilitating Internet use, since they introduced search engines. All users had to do now was go to a search engine site, type what they were looking for, and be provided with a sleuth of information for them to choose from.

The numbers of Internet users exploded at a faster rate than anybody had expected. To give you an idea, it took radio 38 years to have an audience of

50 million people. It took television 13 years to reach the same audience figure. The Internet, once it was open to the public, reached this audience mark of 50 million in a mere 4 years. Today, Internet users worldwide are estimated at more than 1 billion, and are increasing every day (Internet World Stats, 2006). The Internet Society estimated that by 2010, 80% of the planet will have access to the Internet. (See Table 3.1 for an illustration of the growth of Internet hosts worldwide).

Internet technology changes so fast and in most unpredictable ways. When the Internet was first conceived of, it was designed as a system of computer connections to facilitate the sharing among the scientific community of resources such as hardware, software, and data files. Although this model was working, the network remained underutilized for most of its early days. Then unintentionally, the user community came across e-mail as a wonderful and useful application, and the future of the Internet was never going to be the same. People

Table 3.1. Number of Internet hosts by year

Year	Number of Internet hosts
1969	4
1973	40
1977	111
1981	213
1983	562
1984	1,000
1986	5,000
1987	10,000
1989	100,000
1990	300,000
1992	1 million
1993	3 million
1995	7 million
1997	20 million
2001	150 million
2004	300 million
2006 (July)	Over 440 million

Note. Data compiled from Global Reach, 2004; Internet Society, 2006; Internet Systems Consortium (ISC), 2006.

realized that the most powerful aspect of the Internet is its ability to bring together people of similar interests (academic or otherwise) and enable them to become partners in a virtual community, where they could interact freely in real time or at least in a very fast manner.

Today, the Internet is realizing potentials never thought possible. More and more media convergence is allowing for the integration of data, video and audio signals on the network. Broadband technologies, as well as advances in telephone and cable technologies, are allowing new services in a more efficient manner. New technological advances move faster than anyone is able to expect or react to. Newer communication technologies such as wireless technologies, Personal Data Assistants (PDAs), iPods, BlackBerries, and mobile phones are changing the nature of the Internet more and more in a revolution similar to that caused by the introduction of e-mail applications. Such hand-held devices enable ordinary users to check their e-mail and send messages practically anywhere, any time. Users can watch all their favorite videos or listen to their favorite music literally from the palm of their hands. They can call their coffee maker at home to have their drink hot and fresh when they arrive. The possibilities are endless. It is exciting to think of what is possible today, and what will be possible in the short time between this script leaving my hands and until it is printed in book form. As for what will be possible five or ten years from now, I think some of it might be beyond our humble dreams.

· 4 ·

THE INTERNET IN THE ARAB WORLD

When the Internet started out in the early 1960s as an information and communication network for the U.S. army, no one expected that it would have the enormous impact on our lives that it has today. No one expected that it would create such an information revolution, likened only to the invention of the printing press or the telephone, and occupying the minds of the many researchers that it does today.

As discussed earlier, it was only in the early 1990s that the use of the Internet as a public network started to gain momentum. Since then, however, the number of people using the Network of Networks has been on a continuous steep rise, so much so that the latest 2004–2005 Global Information Technology Report (GITR) has declared it a "golden age" for telecommunications and information technologies. The latest estimates assess world Internet users in September 2006 at more than 1.09 billion (Internet Society, 2006; Internet World Stats, 2006; World Economic Forum [WEF], 2005). With that many users and a huge growth rate, this new medium has become impossible to ignore.

The Digital Divide

Equally impossible to ignore are the huge differences in Internet accessibility and knowledge acquisition between the core and periphery countries of the world, a phenomenon referred to as the "digital divide." According to a report by the International Telecommunications Union (ITU), the majority of the world's more than 440 million Internet hosts are in developed countries, suggesting that wealth and education are major factors driving Internet diffusion. The status of a country's socioeconomic development seems to severely affect its chances of being Internet connected. Also of influence are factors such as the price of Internet connectivity, the status of the infrastructure, and language (ITU, 1999; UNDP & AFESD, 2002, 2003, 2004). The UNDP Human Development Report 2001, which was dedicated to the use of technology in development, noted that there were several challenges that face developing nations in terms of acquiring and using technologies, including shortage of trained IT human resources, weak communication strategies, cost, and insufficient feedback mechanisms (UNDP, 2001). At its very basic level, the digital divide refers to differences in user numbers and their access to technology, in other words, the haves and have-nots of information and telecommunications technologies. This, however, is not the only concept that the term refers to. The digital divide can also refer to the quality and speed of connectivity, services available to the user, services actually used, and available information and communication technology (ICT) training or resources. On a more micro level, the digital divide also refers to the haves and have-nots of information technologies within the same country, who are perhaps affected by their race, gender, education, socioeconomic status, or a combination of these factors.

A look at Internet user statistics worldwide as of September 2006 shows how deep this digital divide is. In North America, the United States has a 69.3% penetration rate, and Canada follows closely at 67.9%. The North American continent, which has 5.0% of the world's total population, has an average penetration rate of 69.1% compared with 13.9% for the rest of the world. North America also has 21.1% of world Internet users, while Africa, with 14.1% of the world's population, has only 3.0% of Internet users worldwide, with an average penetration of 3.6%. The top 10 countries in terms of Internet user numbers constitute 62.9% of the world Internet user population. The top 20 countries constitute 78.3% of worldwide users. The top 10 countries in terms of broadband Internet subscribers comprise 80% of the world's fast access Internet users, while the top 20 countries comprise 95% (Internet World Stats, 2006; UNDP, 2005).

These huge differences exist worldwide as well as within each continent. For example, in Europe, the European Union nations represent 22.1% of world Internet users, while the rest of Europe represents 6.3%. Penetration rates differ greatly, with Sweden leading the way at 74.9% in September 2006, followed by Portugal at 74.1%, Denmark at 69.4%, Luxembourg at 68.6%, the Netherlands at 65.9%, the United Kingdom at 62.5%, Finland also at 62.5%, and Germany at 61.3%. On the other hand, countries on the low end of the digital divide have penetration figures such as 2.4% in Albania, 10.6% in Moldova, 11.4% in the Ukraine, and 16.5% in Russia. Overall, there is still a big difference between the average penetration rate in EU countries of 51.9% and that of non-EU countries of 18.7% (Internet World Stats, 2006; UNDP, 2005).

In Asia, the digital divide is even greater. As of September 2006, Hong Kong was leading the continent with a penetration rate of 69.2%, followed by Japan at 67.2%, Singapore also at 67.2%, South Korea at 67.0%, and Taiwan at 60.3%. On the other side of the digital divide is Afghanistan with a penetration rate of 0.1%, East Timor with 0.1%, Myanmar with 0.1%, Bangladesh with 0.2%, Cambodia with 0.3%, Laos with 0.4%, Nepal with 0.7%, and Sri Lanka with 1.4% (Internet World Stats, 2006; UNDP, 2005). While the average Internet penetration rate for the top five Asian countries is 66.2%, the average for the latter seven countries is 0.17%, with Sri Lanka raising that average to 0.36%.

The Arab world is one area that suffers from being on the low end of the digital divide. Out of the 1.09 billion users worldwide, the estimated number of users in the Arab world by September 2006 was around 21 million,[1] less than 2% of the world Internet user population. With some recent developments in the IT sectors of Arab countries, the Internet growth rate in the Arab world is exploding. However, the fact remains that Internet users in the Arab world are about 7% of the Arab population of over 300 million (Abdulla, 2001, 2005; Global Reach, 2004; Internet World Stats, 2006; UNDP, 2005).

One report by the Arab Economic Union Council attempted to specify the reasons for the digital divide in the Arab world. Among the factors cited were the lack of human and economic IT resources, illiteracy and computer illiteracy, the lack of funds for IT research and development, the lack of sc... telecommunication infrastructures in the Arab world, and the cost of ... connectivity in most Arab countries. Another important fact... of an e-business and e-banking culture, and the unav... commerce Web sites and electronic signature vali... Esmat, 2002).

It has to be noted, however, that the worldwide ... their wide variations, actually indicate that the digital

GITR (2003–2004) and (2004–2005), published by the World Economic Forum, estimated that the developing countries were marching fast toward decreasing the gap. According to the reports, out of 200 countries whose data were analyzed, 180 managed to double their Internet users during the years 1999–2002. The reports indicated that most of these countries are developing countries. With respect to the Arab world, the 2003–2004 report cited Egypt and Saudi Arabia as the two countries with the highest rate of IT diffusion. Saudi Arabia managed to increase its Internet users from 100,000 in 1999 to 1.5 million in 2002 to 2.5 million in January 2006. The country had an additional 4.2 million mobile telephone subscribers, and 1.8 million personal computers. The report also called Egypt's performance "remarkable," with an additional 4 million mobile phone users, 2.7 million main telephone lines, 1.4 million television receivers, and 1.3 million Internet users. The number of Internet users in Egypt reached 5.2 million in June 2006 (Internet World Stats, 2006; MCIT, 2006; WEF, 2004, 2005).

Overall, however, the 2003–2004 GITR documented that although the diffusion of ICT has been dramatic worldwide, the largest increases in ICT diffusion came from the most populous countries of the world, including China, India, the United States, Indonesia, and Pakistan. The best Internet user penetration rates, however, are still in advanced countries with smaller populations, such as Iceland, Sweden, the Netherlands, Finland, and Norway. The report also noted that although the most populous countries have had dramatic increases in connectivity and penetration, it is still in those countries that the greatest number of people can still be potentially connected to the Internet. The report concluded that although the digital divide has narrowed, the low income countries are still lagging considerably behind and have much room for improvement (WEF, 2004).

It is estimated that, on average, every Internet account all over the Arab world is shared by about three people, with the exception of Egypt where it is estimated that every Internet account is shared by eight people. (These estimates preceded the free Internet initiative). Estimates for other individual Arab countries include five users per account for Syria, Tunisia, Libya, and Sudan; followed by four users for Yemen; 3.5 users for Lebanon, Jordan and Morocco; three users for Saudi Arabia; and 2.5 users for the rest of the Gulf ates (Abdulla, 2001; Jarrah, 2000). Table 4.1 presents the latest available sta- cs as of September 2006 on numbers of Internet users in the Arab world and, e sake of comparison, some other world countries. The reader has to keep l though that determining the number of Internet users in the Arab major challenge, since all what is available are estimates that vary

Table 4.1. Latest figures available on Internet users in the Arab world and selected world countries as of September 2006

Country	Internet users in 2000	Internet users in 2001	Internet users in 2002	Internet users in 2006	Internet users in 2006 as % of population
Algeria	50,000	180,000	500,000	1.9 million	5.8
Bahrain	40,000	140,200	172,000	153,000	21.1
Egypt	250,000	1 million	3 million	5.2 million	7.0
Iraq	12,500			36,000	0.2
Jordan	127,300	212,000	307,000	629,500	11.9
Kuwait	150,000	200,000	254,400	700,000	26.6
Lebanon	200,000	270,000	432,000	700,000	15.5
Libyan Arab Jamahiriya	7,500	20,000	120,000	205,000	3.3
Morocco	120,000	400,000	722,000	4.6 million	15.2
Oman	90,000	120,000	200,000	245,000	10.5
Occupied Palestinian Territories	23,500	60,000	103,000	243,000	7.5
Qatar	45,000	65,000	69,000	165,000	20.7
Saudi Arabia	300,000	570,000	1.5 million	2.5 million	10.8
Sudan	15,000	56,000	70,000	550,000	1.6

Table 4.1. (*continued*)

Country	Internet users in 2000	Internet users in 2001	Internet users in 2002	Internet users in 2006	Internet users in 2006 as % of population
Syrian Arab Republic	30,000	60,000	226,200	800,000	4.2
Tunisia	110,000	400,000	505,500	953,000	9.3
United Arab Emirates	735,000	880,000	910,000	1.4 million	36.5
Yemen	12,000	17,000	96,500	220,000	1.1
Arab States	**2.3 million**	**5.2 million**	**8 million**	**21.2 million**	**7**
Compare with					
United States	164 million	165 million	166 million	207 million	69.3
Norway	2.2 million	2.46 million	2.68 million	3.14 million	67.8
France	9 million	15 million	19 million	29.5 million	48.4
Sweden	4.5 million	5.74 million	6.02 million	6.8 million	74.9
United Kingdom	20 million	34 million	34.3 million	37.6 million	62.5
World	**451 million**	**680 million**	**800 million**	**1.09 billion**	**16.7**

Note. Data compiled from Global Reach, 2004; MCIT, 2006; UNDP, 2002, 2003, 2004, 2005; US-Egypt President's Council, 2002; WEF, 2005.

sometimes according to the source. The availability of Internet cafes are also another variable that affect an accurate estimate, since most of these estimates are based on computer ownership and Internet subscriptions through Internet service providers (ISPs).

The highest growth rates in the Arab world for the past few years were in Egypt, Algeria, Libya, Syria, Tunisia, and Morocco. Egypt comes first among the Arab countries in terms of the amount of data transferred. The more affluent Gulf states are also growing rapidly in terms of Internet use. Saudi Arabia, a new comer who was introduced to the Internet only in 1999, now has an Internet population of 2.5 million users. The United Arab Emirates, with a population of 3.8 million and an Internet population of about 1.4 million, now has the highest Internet user population percentage of 36.5%. Kuwait follows with a user percentage of 26.6%. Bahrain and Qatar come next with a penetration percentage of 21.1% and 20.7% respectively. Other than Saudi Arabia, late Arab comers to the Internet world include Syria, Libya, and Sudan. While Libya and Syria are moving ahead with high growth rates, Sudan still lags behind with an Internet user population of 550,000 and a penetration rate of 1.6%. It is followed only by Yemen, with a user population of 220,000 and a penetration rate of 1.1% (UNDP, 2005; WEF, 2004, 2005).

Most of the Internet user populations in the Arab world are youths. For example, the Saudi Ministry of Information accounts of Internet users say that over 80% are 20–35 years old, and that males constitute a majority of the users (over 75%). The average time spent on the Internet by any user is 3.5 hours per day. A great majority of users (93%) reported surfing the Web, 72% reported using e-mail, and 32% reported using the Internet for chatting (Saudi Internet Services Unit [ISU], 2004).

Efforts to Increase Internet Penetration in Arab Countries

Most Arab countries are working hard to increase Internet connectivity to their masses. As illustrated in chapter 2, most Arab countries have built good infrastructure for their telecommunications systems, and some are putting in major efforts to spread awareness of the Internet and its potential contributions to the development and well-being of their societies. Kuwait, for example, has rebuilt its telecommunications system from scratch after the 1991 Gulf war. It now has a fully digitized network, and offers advanced Internet services such as fast access Integrated Services Digital Network (ISDN). The fiber-optic network allows for faster Internet access to Kuwait's small population of 2.6 million. The country also has a submarine cable system (FOG) that connects it to other Gulf states. Kuwait's population enjoys a relatively high per capita income compared with the rest of the Arab world. The country also enjoys a high literacy rate of 87%, and a high rate of English speakers, which makes it a potentially heavy Internet user population.

Saudi Arabia, even though it has been very cautious about approaching the Internet, has also made some efforts to increase connectivity, albeit monitored connectivity. As mentioned before, the user numbers have increased drastically in the last four years. The Saudi King Abdel Aziz City for Sciences and Technology (KACST) has conducted two Internet user surveys, which are available on its Web site, to assess usage patterns. Young people, particularly males (75%), constitute the vast majority of Saudi Internet users, with 40% of the user population between the ages of 16 and 25, and another 45% between the ages of 26 and 35. College students constitute 54% of Saudi users. Saudi Arabia enjoys a high level of computer penetration, with only 2% of Internet users saying that they do not have one or more computers at home. Saudi Arabia also introduced fast access Internet, although it is not yet very popular (only 6% reported using it). Most users prefer to connect from home (80%), and use e-mail and Web browsing more than any other Internet application. They seek several types of information and activities online, including scientific research (13%), news (15%), and general information (16%). Chatting about social and professional issues is popular among only 5% of the user population, and entertainment among 9% (Saudi ISU, 2004).

Tunisia is also trying hard to increase Internet penetration, despite close monitoring. Electronic commerce was high on the Tunisian agenda of priorities, so in 1997 the government established the National Commission for Electronic Commerce, a joint effort between the Ministry of Communications and the Ministry of Commerce. As part of the efforts of the commission, more than US$1.5 billion were invested in the Internet and telecommunications infrastructure of the country (Zitoun & Sghaier, 2000). An ambitious plan was put forward to increase Internet penetration, particularly by providing access to schools, universities, and research institutions, as well as by establishing Internet cafes or public Internet centers, called Publinets. One excellent idea that Tunisia implemented was launching what they called "Internet Caravans." An Internet Caravan consisted of several buses, electronically equipped with computers and mobile Internet access. These caravans traveled to various governorates in Tunisia, where they would spend a day of seminars and demonstration workshops on Internet use. The government also implemented "open Internet days," when Publinets became free to the public. These efforts were instrumental in increasing the numbers of Internet users in Tunisia. Indeed, Tunisia was praised by The Internet Corporation for Assigned Names and Numbers (ICANN), the worldwide organization for Internet and Web regulation, for its efforts to increase the user penetration of the Internet. In 2005, Tunisia hosted the second phase of the World Summit on the Information Society (WSIS).

In Jordan, King Abdullah II is very enthusiastic about increasing Internet penetration among the citizens. There has been a National Information Center (NIC) in Jordan since 1993, with the aim of linking public and private information centers. The NIC was the original Internet service provider in Jordan when the service was introduced in 1995, but only within governmental and academic circles. One year later, Internet access became available to the public through several ISPs, who choose their way of accessing the network through any international gateway. Additionally, the government encourages a high level of Internet awareness among the citizens to encourage them to become Internet users. Computers are available in all schools and universities in Jordan, with an estimated penetration rate of one computer per 20 students in public schools (Ein-Dor, Goodman, & Wolcott, 1999).

The United Arab Emirates has also been advocating Internet use among its citizens. As mentioned earlier, the United Arab Emirates now has the highest percentage of population connectivity in the Arab world at 36.5%. The telecommunications system in the country is one of the most sophisticated in the region, being both efficient and cost-effective. Emirates Telecommunications Corporation, *Etisalat*, has maintained a highly developed infrastructure, including a separate submarine cable unit, E-Marine. E-Marine has installed the FLAG (Fibre Optic Link Around the Globe) system all the way to Jeddah in Saudi Arabia, and Aqaba in Jordan. FLAG is an extensive submarine cable system, linking Europe, Asia, and the Far East.

All Arab countries have some governmental representation on the Web, with some being more advanced than others. Saudi Arabia has several well-designed sites, mainly aimed at providing information about the Kingdom to the outside world. The United Arab Emirates has several sophisticated sites, representing all ministries in the country. The country is trying to enhance online governmental communications with the public and also to become a model for Arab world e-commerce efforts. The United Arab Emirates also created an information technology think tank, the Emirates Center for Strategic Studies and Research, with the aim of promoting research in several areas, including the Internet and IT services in the country. Jordan is also well represented in terms of governmental Web sites, with even interactive Web sites, including "Ask the Government" sites. Egypt, as will be explained in more detail in the following chapter, has also made tremendous improvement in this regard. The country has several governmental Web sites, and all ministries are represented on the Web. Recently, in trying to achieve a good model of electronic government, the ministries have updated their Web sites to include up-to-date information about various needs and procedures, as well as contact numbers and e-mails for all departments, options for online communication with the ministries, as well as some e-government applications.

Among the best efforts for increasing Internet connectivity in the Arab world are two Internet cities in Dubai and Cairo. In October 2000, the United Arab Emirates established the first Arab free trade zone dedicated to the Internet, Dubai Internet City. The City, backed by the UAE government, offers a multitude of services and facilities to local, regional, and international companies in the field of ICT. With a total cost of over $200 million, the City aims to attract international key players in IT through tax-free trading. One of the most important features of the City is the inclusion of research and development units in IT. The City also has links to media outlets and cultural projects in an attempt to bring together information, media, and culture. It is managed by a group of young experts in IT and business management. The City occupies 37,000 square meters. When it was inaugurated in 2000, about 200 companies had already bought or rented space within the City. Today, Dubai Internet City hosts many technology giants, including Microsoft, AMD, Oracle, HP, IBM, Compaq, Dell, Siemens, and Sony Ericsson. (Dubai Internet City [DIC], 2006; UNDP & AFESD, 2002).

Egypt also has established a city for the conglomeration of IT and communications companies. Egypt's Smart Village is a 300-acre project with 58 office plots, accommodating about 30,000 employees within a total office area of 1,336,000 square meters. This office area is only 10% of the total area of the village, with the remaining space dedicated to green areas, lakes and water streams. Like Dubai Internet City, Egypt's Smart Village is also backed by the government, with the Egyptian Ministry of Communications and Information Technology moving there in 2004. Investors can buy or rent land and office space, and are given a ten-year tax exemption. The Village is equipped with state-of-the-art telecommunications and technology facilities, and hosts a plethora of technology companies, including Alcatel, Microsoft, HP, Vodafone, MobiNil, IBM, Canon, Orascom Technologies, and LINKdotNET (Egypt Smart Village [ESV], 2006) (Figure 4.1).

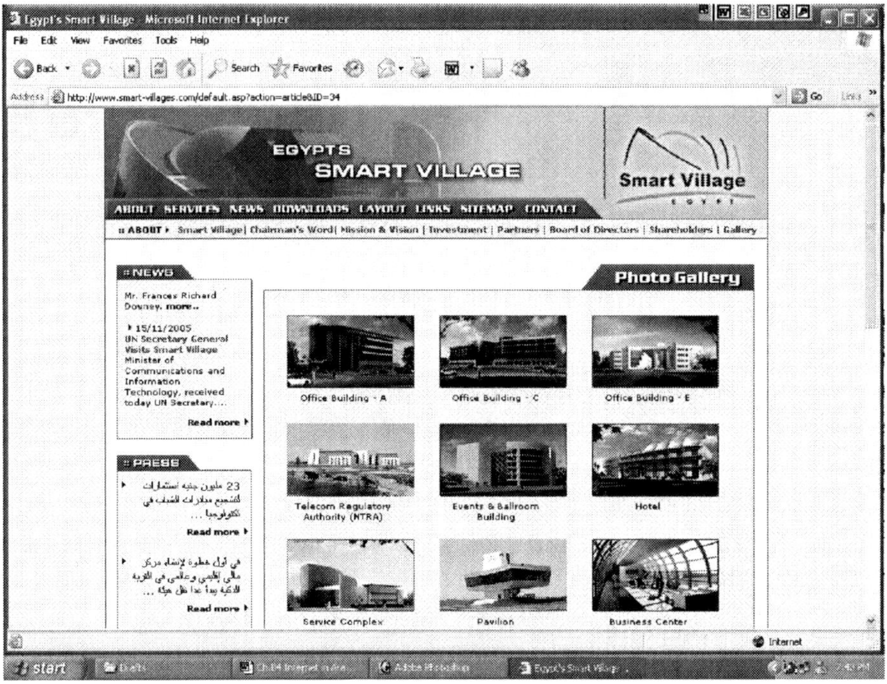

Figure 4.1. Egypt's Smart Village Web site

Among the main problems that Internet users in the Arab world face is a relative lack of Arab and/or Arabic-language content. While English-language content on the Internet accounted for 28.7% in September 2006, Arabic-language content accounted for less than 2% (Internet World Stats, 2006). However, Arabic-language Uniform Resource Locators (URLs) are starting to surface. About 75,000 Web sites had registered an Arabic-character URL address by July 2001. Several technologies are now available to bypass the Latin character-based Internet domain name system (Dabbagh Information Technology [DIT], 2001). Pyramid Research concluded that "the growing prevalence of Arabic character domain names stand to boost Internet adoption across the Middle East and North Africa. Viewed in a broader context, the 'Arabisation' [sic] of the Web content and computer interface are major drivers of Internet uptake." Accordingly, Pyramid Research expected that new Arab users would soon join in the Internet world, mainly from the middle and lower-middle income classes. These users would not necessarily be fluent in English, but would have greater access to computers. Most of these users will be from the more affluent Gulf States, where English-language proficiency is not necessarily a prerequisite to the maintenance of a comfortable standard of living. These users would have the economic means to afford Internet connectivity, but would have been deterred by a relative lack of Arabic language content. Pyramid Research had projected that the market for Arabic-language URLs and domain names within the Arab world would in a few years be worth US$126 million. Forty-five percent of the demand would come from Egypt, and 43% from the Gulf states, led by Saudi Arabia and the United Arab Emirates (DIT, 2001).

Some other efforts at Arabizing Internet content include translating English Web pages. Some leading ISPs, such as LINKdotNET, are advertising a translation service if users access the Internet through their numbers. Users access the Web site of their choice and click on a link to see the page in Arabic. This service is still in the beginning stages, but so far is proving to be popular among users.

Some Arab countries are also trying to implement ambitious plans to increase computer ownership and penetration. In Egypt, for example, a project was launched in 2002 that aims at having a computer in every household. The project gives the right for anyone who has a fixed telephone landline to buy a personal computer and pay for it in relaxed, no-interest, monthly installments that are automatically added to the phone bill. In 2004, the MCIT undertook an aggressive advertising campaign to market the plan, particularly among the lower classes and citizens of rural areas. The objective of the project is to reach a quota of 100,000 PCs sold annually (MCIT, 2004).

Arab states are also beginning to realize the necessity of integrating computers and Internet access at various educational levels. So far, such access has been deterred by economic difficulties, teacher-training problems, as well as language barriers. However, countries such as Egypt, the United Arab Emirates, Jordan, and Saudi Arabia are now implementing plans to have computer and Internet laboratories in every school. In Jordan, all public schools have computers available to their students at a ratio of one computer per 20 students. Private schools have even better ratios, and all have Internet access. A plan was initiated in 2000 to provide Internet access to all public schools as well.

· 5 ·

THE INTERNET IN EGYPT

Egypt is given special emphasis in this book for several reasons. First, it is the country with the biggest number of Internet users in the Arab world. It is also the country with the biggest amount of data transferred across the Internet. Second, it is a country of more than 77 million people, by far the largest population in the Arab world. It therefore has the most potential for growth in terms of Internet usage. And indeed, the growth rate in Egypt for the past three years has been an annual 500%. Third, as illustrated in chapter 1, Egypt has traditionally been a leader in providing media content in the Arab world. As such, and with a major training program in IT underway in the country, Egypt is in a unique position to provide Internet content that is tailored for the Arab population, and therefore help Internet adoption and diffusion in the region. Fourth, the study carried out by the author, which will be discussed later in the book, targeted Arab students in Egypt, and therefore it is useful to explore the status of the Internet in the country to get a better understanding of the circumstances under which the study was conducted.

As mentioned above, Egypt has one of the largest growth rates of Internet connectivity in the Middle East, currently estimated at 500%. While the number of users was estimated at 220,000 in 1999, it has grown to 5.2 million in September 2006 (Abdulla, 2001; MCIT, 2006). Despite a relatively weak economic situation, the country has spared no effort over the years to increase Internet adoption.

The Internet was introduced to Egypt in October 1993 through the Egyptian Universities Network (EUN) of The Supreme Council of Universities and the Cabinet Information and Decision Support Center (IDSC)'s Regional Information Technology and Software Engineering Center (RITSEC) (EUN, 2001; RITSEC, 2001). Egypt Telecom provided the basic infrastructure for the initial connection, a 9.6 Kbps link to France. This link carried both Internet traffic as well as BITNet, an earlier form of electronic mail communication. The user community was estimated at about 2,000 to 3,000 users, out of Egypt's overall population of 63 million at that time (Internet Society of Egypt, 2001).

In 1994, the Egyptian Internet traffic was divided into three major subdomains: academic, science, and government. The academic subdomain provided free service for schools and universities. It was given the domain name ".eun," which stands for "Egyptian Universities Network." Service to this subdomain was provided through a gateway at the Supreme Council of Universities, and was available only to faculty members and to some senior administrators. Scientific research institutions were served by the ".sci" subdomain. This subdomain was being served for free through ENSTINET (The Egyptian National Scientific and Technical Information Network), as well as other research centers like the National Telecommunication Institute. The government subdomain ".gov.eg" was again provided, free of charge, through IDSC/RITSEC, and served various governmental agencies (Internet Society of Egypt, 2001). At that point, the Internet was not available for commercial purposes in Egypt. There was only one ISP, In-Touch, who managed to get an international connection through a server in the United States. In-Touch could only provide e-mail services at that time. To get an e-mail account at In-Touch, a user had to pay a one-time fee of L.E. 300 (about US$100 given the exchange rate at that time), plus L.E. 2 (US$0.67) for every e-mail sent, L.E. 0.85 (US$0.28) for every 1,000 characters sent or received, and L.E. 10.8 (US$3.6) for every hour connected (Abdulla, 1995).

The original slow connection did not make the Internet attractive enough to potential users, who were going to have to invest time and money to learn and acquire the new technology. However, in September 1994, Egypt hosted the International Conference on Population and Development, an event that had more than 15,000 participants. To host the meeting, Egypt had to provide a 64 Kbps Internet connection to serve the press and organizers. A link was established to France, costing the Egyptian government $450,000 for nine days. After the conference was over, the government decided to keep the line, and offered it to the IDSC/RITSEC. The line was to be used to promote Internet service by allowing further free connectivity to nongovernmental organizations (NGOs) and private businesses[1] (ITU, 2001).

In 1995, IDSC/RITSEC launched the Egypt Information Highway Project to motivate and support the development of a national information highway that would empower Egypt's socio-economic growth. The objectives of the project were:

- To support the information society in Egypt, and help integrate it with the global information society.
- To promote and support electronic dissemination of information over high speed networks.
- To establish pilot information highways in critical areas to energize socio-economic development.
- To contribute towards open and wide access to the national information highway.
- To encourage secure on-line databases to support socio-economic development.
- To assist human resource development required for establishing the national information highway (RITSEC, 2001).

To fulfill those goals, an ambitious plan was put in action to provide digital data access to the end user through digital multiplexers. The first digital public backbone network installation was in Cairo, a joint effort between IDSC/RITSEC and ARENTO. Fiber-optic international connectivity was made available on satellite via Intelsat, thus overcoming the basic obstacles of the infrastructure limitations. Another project aimed at deploying VSAT services for Internet connectivity in Egypt. This provided rural areas in Egypt with the necessary communication infrastructure (Internet Society of Egypt, 2001; RITSEC, 2001).

By January 1996, the gateway speeds were increased by nearly 20 times from their inception in 1993. User numbers increased to 20,000. Private ISPs started operating either through IDSC/RITSEC under the ".com.eg" domain or through their own international gateways. The service therefore became available to the general public. By that time there were three main sources of Internet connectivity in Egypt. The academic and scientific sectors were served by EUN, providing free Internet services to faculty members, staff, and graduate students. IDSC/RITSEC provided free service to government agencies, NGOs, and selected commercial businesses. The third main source of Internet connectivity in the country comprised 12 ISPs, providing full for-fee services to anyone who could afford them. At that time, prices ranged from a thousand to several thousand Egyptian pounds per year (about US$300–1,000 given the exchange rate at that time) (Abdulla, 1995; Internet Society of Egypt, 2001). By 1998, prices ranged from about L.E. 20 to 100 per month (US$5–25). It should be noted that telephony constituted an additional fee for Egyptian Internet users, since telephone communication fees are calculated on a per-minute basis rather than a flat fee.

In January 2002, the Egyptian government embarked on one of the most ambitious initiatives in the Arab world to increase Internet connectivity and user access. The government started a unique service, whereby Internet connectivity, through any of the 68 ISPs then available on the market, became free for all. All that the users had to pay then was the negligible price of the telephone call while being connected to the Internet,[2] a price which they were already paying anyway. Telecom Egypt would then pay 70% of the telephone revenues from Internet use to the respective ISPs. When the government first announced this plan in 2000, it caused drastic price drops in Internet connectivity. Most ISPs made offers for lifetime service for about L.E. 100–200 ($25–50 at that time), in an attempt to lure customers into continuing to dial in through their access numbers once the service became free. The ISPs competed to offer more access lines, free e-mail accounts, and better customer service. Others moved quickly to introduce fast-access broadband Internet service so they can have a competitive edge over the free service. Integrated Services Digital Network (ISDN) Internet access is now available in Egypt for an average of L.E. 450 (US$75) per year for a 64 Kbps connection, or L.E. 700 (US$115) for a 128 Kbps connection. Other options include DSL (Digital Subscriber Line), which is now available for L.E. 150 (US$25) per month, ADSL (Asynchronous DSL), T1, and T3 connections. As of November 2002, the free Internet dial-up service was available in all of Egypt's 26 governorates (US-Egypt President's Council, 2002). The latest available information on Egypt's Internet profile is presented in Table 5.1.

The largest ISP in Egypt is LINKdotNET, the majority shares of which are owned by Orascom Telecom, which serves more than 20 markets in the Arab world and Africa. The company offers Internet access through a range of connectivity and speed options (including free Internet access), Web page design, Web content development and hosting services, e-commerce and e-business services, online advertising, as well as wireless applications services. LINKdotNET was created as a result of a merger between Egypt's first and largest two ISPs, Link Egypt and In-Touch Communications. In May 2002, the company acquired eight other major Egyptian Internet companies,

Table 5.1. Egypt's Internet profile as of September 2006

Number of users	5.2 million
Users annual growth rate (2002–2005)	500%
Bandwidth	5.355 Gbps
Number of ISPs	300
Registered IT clubs	1,042
IT companies	1,494
Communications companies	76
IT registered employees	36,394
Communications registered employees	5,478
Trainees in IT	149,480
Annual market growth rate	25%

Note. Data compiled from MCIT, 2006.

significantly raising the company's value, and making it the largest Internet company in the Arab world. Currently, LINKdotNET has about 400,000 subscribers in Egypt alone, served by some 600 professionals. The company maintains branch offices in Saudi Arabia and the United Arab Emirates to serve its businesses in the Gulf region. It is also a partner in MSN Arabia, one of the major portals serving the Arab world both in Arabic and in English (LINKdotNET, 2006).

There is currently a huge number of Internet cafes in Egypt, even in the most rural and the poorest areas throughout the country. One report (Human Rights Watch [HRW], 2005) indicated that in 2005, there were 460 Internet cafes in one small city of Sharkiya governorate, mostly operated by students of that city's local university.

Egypt led the Arab world in its growing efforts to Arabize Internet content, when Egypt's Information Highway Project put the first Arabic Web pages on the Net in 1996. The Egyptian *Al Gomhuria*[3] was the first Arab newspaper to launch an electronic version in 1996. *Al Ahram*[4] newspaper, Egypt's most widely distributed daily since 1876, launched an electronic version of its content in 1998. At that time, most browsers did not support the Arabic language, so Arabic content had to be uploaded as pictures, causing problems such as a slow speed of surfing and downloading information (Nagib, 2000). This problem was solved with the introduction of the following generation of Web browsers, which supported Arabic text. Arabic newspapers were further able to index news stories using keywords for search purposes, which made the sites more appealing for both journalists and readers. Today, most Arabic- and English-language Egyptian national publications (at least 40) have electronic versions on the Web. One Web site, *Sahafa*[5] Online, lists at least 250 Arab online publications.

There are no definite figures about the demographics of Internet users in Egypt. Estimates, however, predict that more than 40% of the population of Internet users in Egypt is comprised of students aged 16–28 (A. El Messidi,[6] personal communication, December 28, 2001). One study estimates the majority of users (53%) to be in the age group of 21–29 years, with another 24% between 30 and 39 years of age, and 15% to be between 40 and 45 years old. Most users (88.5%) have some university education (Hamdy, 2001).

Egypt has been hosting several annual international conventions on information technology and the Internet, including CAINet, Cairo Telecom, Comdex, Gitex, and Cairo ICT since 1996. The conventions have more than 5,000 participants each, and are held under the auspices of MCIT, who also organized Africa Telecom in May 2004, and a pan-Arab conference in preparation for the WSIS in May 2005. An Egyptian delegation attends all ICANN meetings, and a high-level delegation attended the WSIS in Tunisia in November 2005.

One of the more successful efforts at spreading Internet awareness and education has been carried out by UNDP in Egypt. Technology Access Community Centers (TACCs) were launched in March 1999 in two cities in Sharkiya Governorate, some 80 kilometers north of Cairo. The project represents a "cybercafe for the poor." It aims at providing affordable public access to the Internet and other communication technologies to residents of rural and remote communities. The project also provides free training on the use of such technologies. A TACC's location should be central to the community it serves. It includes not only Internet-enabled computers, but also fax machines, copiers, printers, and software libraries. Training, seminars, and workshops are provided by specialized personnel in an effort to empower the local community. Through educating customers on how to access information, and how to develop Web pages of local interest to the users, the project helps the development of a sense of community as well. In its pilot stages, 6,742 users benefited from TACCs. More than 1,400 Web pages were developed in Arabic, including a much-needed database of local medical practitioners that is now hosted and maintained by the Ministry of Health and Population Web site. The project also introduced the concept of e-commerce in Sharkiya

Governorate, and encouraged the establishment of IT-related businesses, cyber cafes, and Web development projects. The success of the project encouraged its replication in two other Egyptian areas: Siwa, and Luxor. It is estimated that each TACC is visited by an average of 600 users per month. On the Arab level, replication is also underway in Jordan, Syria, and Yemen (TACC, 2002).

There are now more than 1,000 information technology clubs registered in Egypt. Among them are what is known as the 21st Century Kids Clubs, which were established as a joint effort between the government, NGOs, and the private sector. First established in 1997, these clubs provide Internet access, software, and ICT trainers to help children navigate their way through technology. The clubs are managed by Integrated Care Society, an NGO, and are part of a RITSEC initiative to spread ICT awareness among children. In 2001, there were 40 kid clubs available in 17 governorates. Today, there are 400 clubs in Egypt's 26 governorates. The goal is to establish over 1,000 clubs all over Egypt. To serve the 21st Century Kids Clubs, RITSEC launched the first official children's Web site, Little Horus,[7] in 1997. The site, which can be accessed in both Arabic and English, contains information for children about Egypt's history, traditions, children, and places to visit. It also includes links to children's clubs worldwide, and encourages Egyptian kids to interact with others and to "make one friend a day." The site was popular among kids and received several awards in 1999 including the Internet Society of Egypt Award, the CAINet '99 Conference and Exhibition Award, and the Cable and Wireless Childnet International Award (Little Horus, 2006; MCIT, 2006; School Net Africa [SNA], 2003) (Figure 5.1).

Internet regulation in Egypt is carried out by the National Telecommunication Regulatory Authority (NTRA) under the auspices of MCIT. The goal is to improve the legal and regulatory framework for the Internet and IT sector in Egypt, as well as to expand adoption and delivery of the Internet and related applications in the country. The NTRA was established as per the 2003 Telecommunications Regulation Act (law No. 10/2003), which was passed by the Egyptian Parliament to outline the general rules guiding all types of telecommunications operations in Egypt. The NTRA is responsible for all licenses and permits pertaining to telecommunications in Egypt, as well as the regulation and pricing of the Egyptian frequency spectrum or bandwidth usage (NTRA, 2006).

In 2004, the Egyptian Parliament voted to pass the Law Regulating Electronic Signatures, also known as the E-Signature Law (No. 15/2004), which aims to build a proper infrastructure for e-business and e-commerce in Egypt. The law

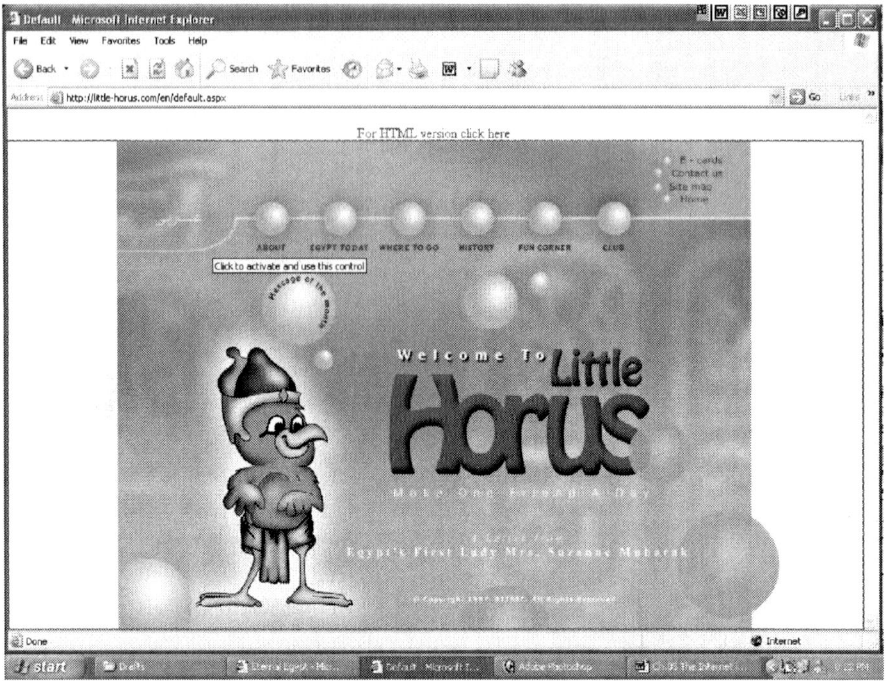

Figure 5.1. Egypt's Little Horus Web site.

established an e-signature regulatory body, known as the Information Technology Industry Development Agency (ITIDA), which also functions under the auspices of MCIT. This agency is responsible for issuing digital certificates to a group of Certificate Service Providers (CSPs). The ITIDA is responsible for monitoring and auditing the security systems of CSPs in order to protect individual customers as well as companies and corporations using the systems. The whole e-signature system should be fully operational by the end of 2006. The ITIDA has also set up an ICT industry ombudsman, which will act as a mediating body in case of disputes (ITIDA, 2006).

Other efforts for establishing e-commerce and e-business have started to emerge in the Egyptian society, albeit at a slow pace. These efforts include establishing Web sites for major Egyptian financial institutions, including the Central Bank of Egypt (www.cbe.org.eg), and the National Bank of Egypt (www.nbe.org.eg). These sites are mostly informational in nature, and have not yet allowed e-banking transactions. However, ten other major banks have started offering online banking and telebanking services in Egypt. Banks have started aggressive marketing campaigns to spread the adoption and use of credit cards, which is currently rarely used in this cash-oriented society. A project to further spread the use of Asynchronous Transfer Mode (ATM) networks, and therefore encourage the use of banking cards, has been underway with a budget of US$600 million. As a result, a growing number of Web sites are starting to offer services ranging from selling cars, to real estate, to food and services delivery, to hotel and car rental reservations (through e-mail or pay-upon-delivery).

Other sites try to find creative ways to answer customer demands and get over e-commerce obstacles. For example, www.otlob.com[8] allows users to place orders online for a variety of food, pharmaceuticals, flowers, video rentals, electronics, and even mobile phones and computers. Orders for most goods are delivered to the customer's address within half an hour to an hour. The site offers access to menus and prices of over 500 restaurants and a large number of companies and different service providers. Available for Cairo and Alexandria users, the site still does not employ the use of credit cards. Rather, it offers a cash-upon-delivery system.

The lack of a banking card culture is currently high on the list of obstacles to e-commerce in Egypt from a consumer's standpoint. Although there are over 100 banks in Egypt, there were less than half a million cards in use in 2001, less than 0.75% of the population (ITU, 2001). To obtain a credit card in Egypt, customers on average have to maintain a deposit of 150% of their credit limit, which poses a problem to most users. Other social obstacles include a lack of awareness of the potentials of e-commerce, and a lack of trust in electronic payment systems. Resisting change is another important factor in a culture where commercial life has depended on cash for so long, which is what the government and the various financial institutions are trying to change.

In this regard, the Egyptian government has been working on encouraging the integration of e-commerce and e-business options into the daily lives of Egyptian Internet users. Areas being looked at include laws and regulations, telecommunications infrastructure, and payment security issues. The Electronic Commerce Committee of the Internet Society of Egypt has been organizing workshops and seminars to spread awareness of the potential of e-commerce. Such potential is especially evident in sectors such as tourism and trade, where Egypt's primary business partners, namely, in the United States and Europe, already have the means (in terms of Internet access, e-commerce systems, and economic affordability) to conduct transactions.

One of Egypt's most recent ICT projects is the establishment of call centers that serve major clients around the world. Through outsourcing their call center services to Egypt, these clients get a friendly, highly educated team of operators,[9] who are available to answer calls in Arabic, English, French, German, Spanish, Italian, and Russian. The centers use a number of global and regional fiber-optic cables that provide reliable international connectivity. A recent development was the authorization of Voice over Internet Protocol (VoIP) for use by these call centers, thus enabling them to provide their services at competitive prices, and fulfilling an international connectivity requirement for the call center industry. With Egypt strategically located at the crossroads of three continents, the government expects this service to boom in the next few years (Call Centers Egypt, 2006).

Efforts at developing content have been underway for a while in Egypt. One example is the Information Highway Project, which started developing content in several subsectors, including cultural heritage, tourism, health care, trade, and electronic commerce, as well as government services and public administration. Several information networks were launched including: Health Net, Tourism Net (which is reported to generate more than 60% of the traffic on the main Web site), Environment Net, Culture Net, Governorates Net, Libraries Net, and Government Online. The project's homepage[10] was reported as early as 1997 to have more than 38 million hits from more than 900,000 different IP addresses outside Egypt, more than 10,000 IP addresses in Egypt, and more than 2000 e-mails (Egypt Information Highway, 2001; RITSEC, 2001).

Egypt has taken productive steps toward developing the services of its e-government. Online government agencies can be accessed through the government's portal,[11] which is the result of a joint project between the government and Microsoft, and was inaugurated by Bill Gates on his first ever visit to the Arab world in January 2004. All 26 governorates can be accessed through the site, as can be all 35 national ministries, the Office of the Prime Minister, the Parliament, and the Office of the President. Some e-government services are also fully available online for citizens, with the option of paying for them with a credit card or cash-on-delivery. These services include requesting a birth certificate, requesting a national identification replacement card, automobile license renewals, paying or disputing traffic tickets, checking and paying phone and electricity bills, taxation services, customs services, postal services, tourist complaints, as well as applying for schools and universities.

Other government agencies offer access to their forms and applications, and instructions for obtaining their services. Most of these other sites are informational. While a user can access the required service, get information on the documents needed, fees, and time schedules, and download forms, the user for the most part cannot yet fill out and process the forms online. It has to be noted, however, that the availability of the information and instructions online is a great service to the Egyptian people, who, in this bureaucratic society, otherwise could waste days or months trying to find out what documents are required to process a service. The government aims at offering most of its services online by 2007.

While all ministries are represented by their own Web site, some Web sites are more developed than others. Among the highly developed sites are the MCIT Web site, as well as sites for the ministries of health and population, education, foreign affairs, media affairs, culture, foreign trade, and tourism. For example, the MCIT site is one of the few where users can process information and pay fees online. Customers can access their phone bills, and pay their bills online with a credit card. The site provides access to updated research and statistics pertaining to the status of telecommunications and ITs in Egypt. The ministry also has an online telephone directory[12] of all residential, governmental, and business phone lines in Egypt. The site can be accessed in Arabic or in English, and also includes a directory of hundreds of Egyptian business and governmental Web sites. A CD-ROM of the national telephone directory can be purchased from any central office for L.E. 16.5 (US$2.8). The CD-ROM can also be ordered online, or through a toll-free number (Egyptian Government Web site, 2006; MCIT, 2006) (Figure 5.2).

The Ministry of Health and Population (MOHP) Web site offers a physician's search by name and/or specialization, information on drugs, pharmaceutical companies, hospitals, medical centers, pharmacies, medical societies, as well as links to other valuable medical sites. The site also has a host of information on different health issues, currently foremost among which is the bird flu. The MOHP has recently started a national campaign to register Egyptian diabetes patients and their medical history in an online database that would be available to their respective health practitioners. The project is expected to be replicated for patients of other chronic diseases (MOHP, 2006).

Tourism is one of the areas that can benefit the most from the Internet in Egypt. With the country housing one-third of the world's antiquities, and bordering both the Red Sea and the Mediterranean, Egypt depends on tourism as its number one source of foreign currency. The country is visited by an average of 8 million tourists a year, yielding a revenue of about US$6 billion. The Ministry of Tourism has several tourist Web sites,[13] and uses the Internet, mainly e-mail, to promote tourism through tour operators in the region. The IDSC Information Highway Project also launched Tourism Net,[14] a 25,000-page Web site. Complete with a hieroglyphics translator and a hieroglyphics message

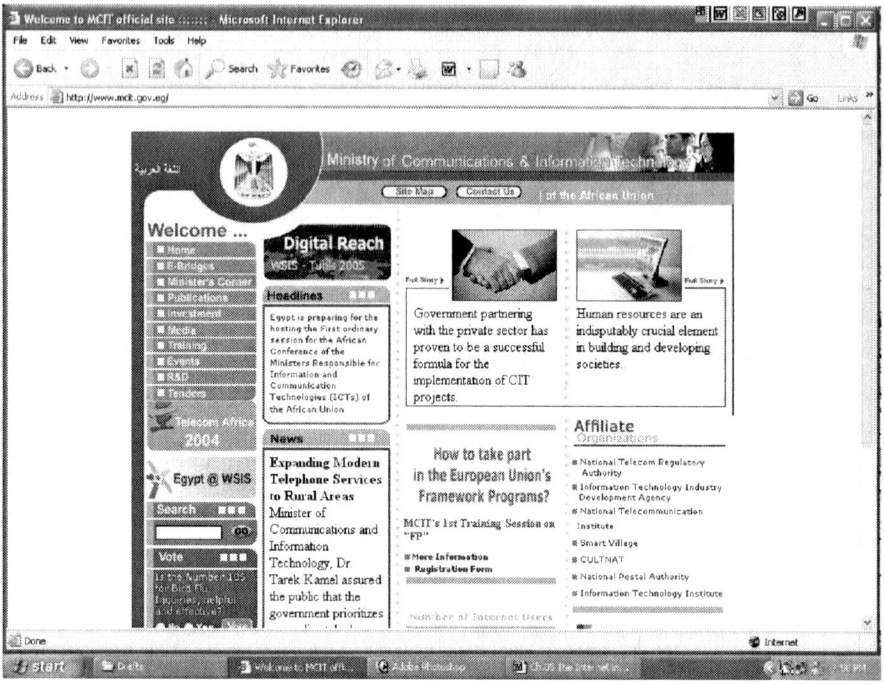

Figure 5.2. Egypt's MCIT Web site.

sender, Tourism Net offers information on all aspects of tourism in Egypt in four languages: Arabic, English, French, and German. The main weakness of these sites again is their being purely informational. Although they do offer a list of addresses, phone numbers, and e-mail addresses for hotels, through which reservations could be made, online reservations are still not available except through foreign international Web sites (Figure 5.3).

Another excellent tourist site is Eternal Egypt,[15] the result of several years of hard work on the part of the Center for Documentation of Cultural and Natural Heritage (CultNat), a center affiliated with Bibliotheca Alexandrina and supported by MCIT (Figure 5.4). With the help of a US$2.5 million grant of technology and expertise from IBM, CultNat worked on documenting every single piece of archeology that is registered in Egypt in three languages: Arabic, English, and French. Over a period of three years, CultNat and IBM then worked on creating high resolution images and three-dimensional, multimedia displays of these Egyptian treasures. The result of this cooperation was Eternal Egypt, which allows its visitors, for example, to enter a three-dimensional reconstruction of Tutankhamun's tomb, as well as of thousands of other arche-

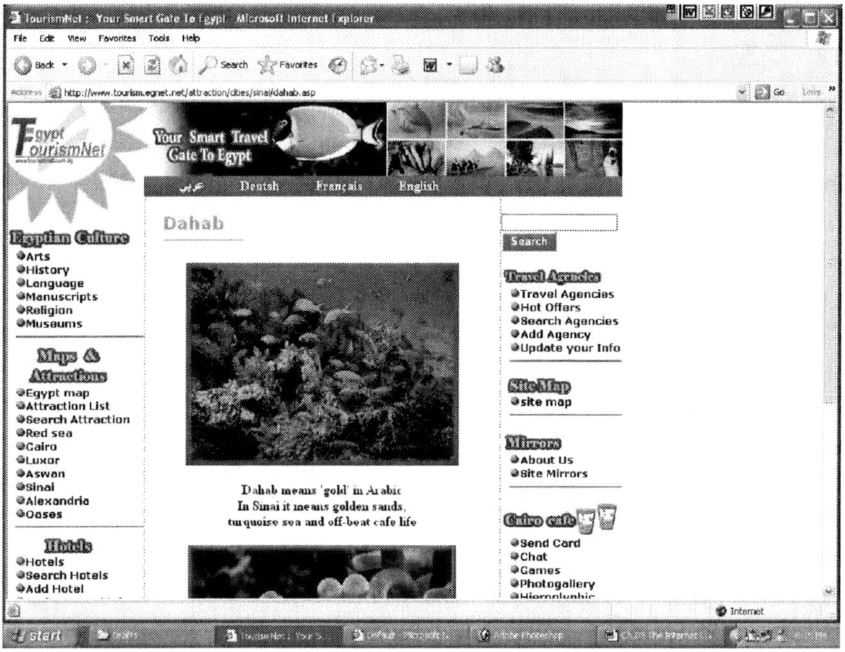

Figure 5.3. Egypt's Tourism Net Web site.

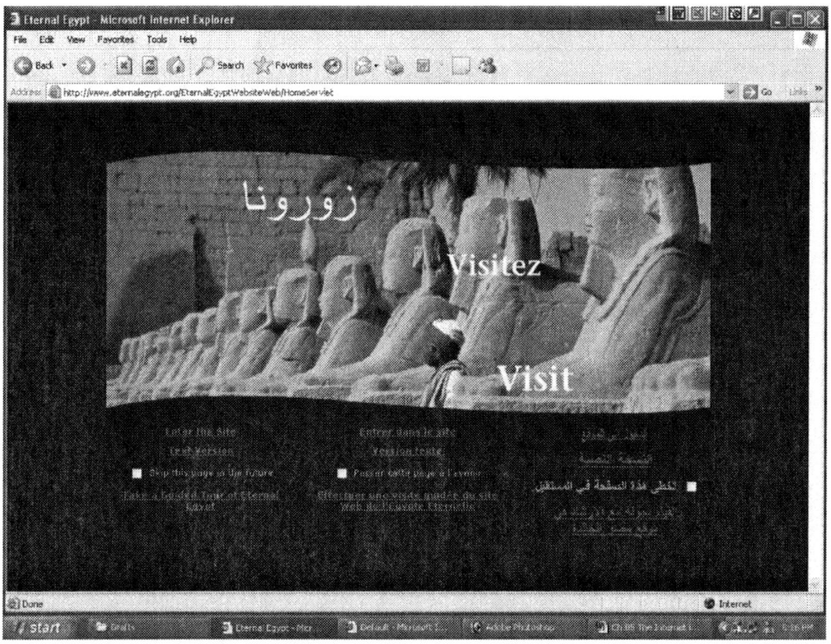

Figure 5.4 Eternal Egypt Web site.

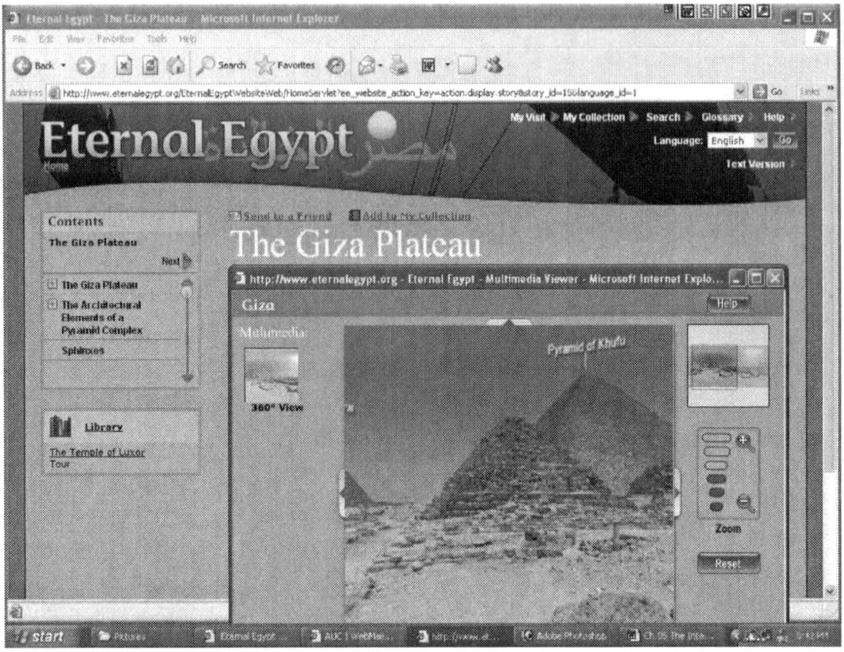

Figure 5.5 A 360-degrees view of the Giza Plateau on Eternal Egypt.

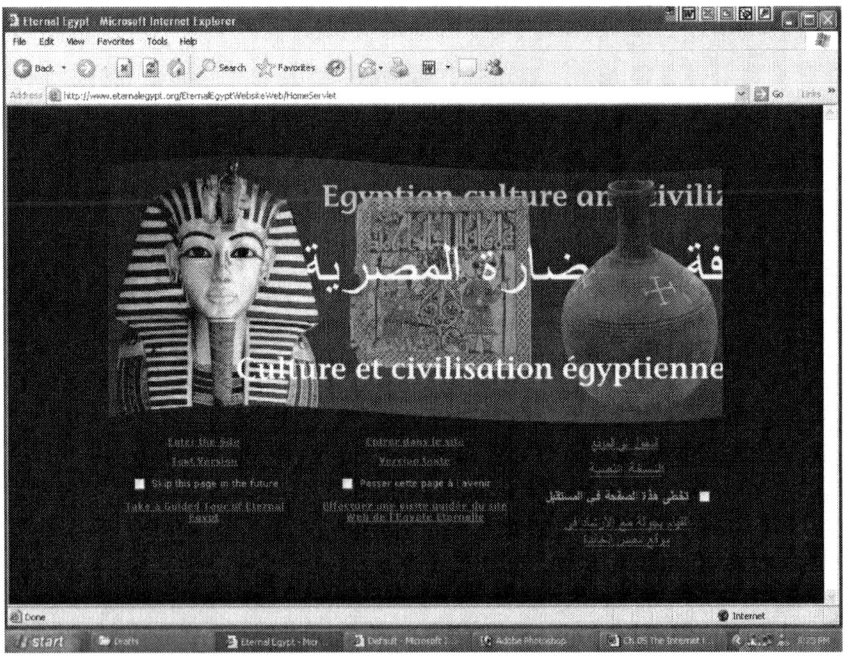

Figure 5.6 Eternal Egypt Web site.

ological sites (Figure 5.5). Eternal Egypt is the only Web site in the world so far that includes a digital guide for mobile access. This service allows users with a data-capable and networked mobile phone anywhere in the world to download all the contents of the Web site, including information, images, videos, and multimedia presentations, through their handheld digital device. It also allows users with a mobile phone or a Personal Digital Assistant (PDA) to download multimedia guided tours if they are in the Egyptian Museum or on the Giza plateau (the Pyramids area) in Cairo or in Luxor. The information is downloadable in any of the three languages available, complete with images reformatted for small screens and an optional audio guide. Eternal Egypt is a highly popular site with lots of awards to its credit, the latest of which is the 2005 WSIS Award for e-culture (Figure 5.6).

With such huge efforts taking place, the information technology sector is currently the largest growing sector of the Egyptian economy, with a value of US$1.3 billion (HRW, 2005; MCIT, 2006).

· 6 ·

ISLAM AND THE INTERNET

A simple search of the word "Islam" on Yahoo! generated 97,300,000 results. Out of curiosity, I decided to compare this result to "Christianity" and "Judaism," those being the three monotheistic religions. I got 58,000,000 and 20,400,000, respectively. Quite intriguing!

Islam is the youngest of the three religions, having started in the year 611 AD. It is also the fastest growing religion, and it is the most controversial of the three religions, particularly in light of the attention it has been given since the attacks of September 11, 2001 on the United States. Since then, it has become commonplace to hear the words "Bin Laden," "terrorism," and "Islam" in the same sentence, much more so than you would hear "McVeigh," "terrorism," and "Christianity" or "Ku Klux Klan," "terrorism," and "Christianity" or "the Stern Gang," "terrorism," and "Judaism."

With such a tremendous amount of Web sites, has the Internet changed anything about Islam and Muslims? Is there a change in the pattern of communication, at least religious communication, amongst the 1.3 billion Muslims that inhabit this planet? This chapter aims to investigate these questions through an exploratory non-random content analysis, scanning the major Islamic Web sites on the Internet, in an attempt to find out what content is offered on these Web sites, the quality and credibility of the information offered, and how Muslims are using the Internet to fulfill their communication needs. I believe that the Internet is changing the fabric of Muslim communication patterns and that at this stage, the Internet is likely to be of more benefit to Muslims living in Western countries than their counterparts in Muslim countries.

Arabs versus Muslims

Since we've been discussing the Arab world in this book, let me quickly clarify the difference between Arabs and Muslims. It is a question of ethnicity versus religion. Islam is not an ethnicity, but a religion. It is the majority religion in the Arab world, with over 95% of Arabs subscribing to the religion. However, the majority of Muslims are not Arabs. Arabs constitute only about 23% of Muslims worldwide. The majority of Muslims come from non-Arab countries such as Indonesia, Malaysia, Afghanistan, Pakistan, and India. It also has to be noted that the 1.3 billion Muslims in the world, although having a common framework of the Quran, do not necessarily share the same views, patterns of life, clothing or life style, or even interpretations of the Quran. Just like Arabs differ, sometimes tremendously, from one country to another, so do Muslims. Generalizations on Arabs, Muslims, or any other group of people for that matter, are gross exaggerations.

Credibility of Islamic Web sites

Having pointed that out, the plethora of Muslim-related Web sites on the Internet includes everything from educational pages to religious portals to shopping sites. It also includes a huge amount of "Islamic" sites designed by non-Muslims, some of which explain their own experiences with the religion, and some provide their interpretations, rightly or wrongly, with what they believe the religion is all about. The topic of "Islam and the Internet" is difficult to tackle in a single book chapter. Nevertheless, I will try to assess some potential effects of the Internet on Islam and Muslim communities, and then I will try to provide some examples of popular Islamic Web sites that serve some basic functions for Muslims and non-Muslims worldwide.

Bunt (2003) noted that Islam is not dependent on computers, does not need the Internet, and probably would not miss it if it disappears. However, he still wrote several books on the subject, reiterating that the Internet has expanded several aspects of Islam. Bunt argues that Islamic propagation should provide an incentive to Internet use among Muslims, which he admits is currently inhibited by the aspects of digital divide earlier discussed in this book, namely, lack of access, cost, illiteracy, computer use, and language.

The number and variety of Islamic Web sites have increased tremendously following the events of September 11, 2001. While most Islamic Web sites tend to be in Arabic (J. Hassan, 2002), some of the more sophisticated ones are bilingual, offering their content in both Arabic and English. These sites contain everything from Quranic texts and recitations, to interpretations, to Islamic *fatwas* (religious, theological, and legal opinions of Islamic scholars), to Friday *Khutab* (sermons), to discussion boards, and frequently asked questions.

However, with the Internet allowing anybody to become a publisher, the problem arises of "unofficial" sites that provide interpretations and *fatwas* that might seem genuine, when in fact very few Muslim scholars would agree with their content. Such materials provide an inaccurate image of the religion that feeds the controversy and the stereotyping surrounding it. According to Khaled Abou el Fadl, professor of Islamic law at UCLA's Center of Near Eastern Studies, the Internet introduces a challenge of legitimacy and accountability assessment for users. In his words, "There are thousands of cyber *Muftis* (religious scholars) out there . . . spewing out these *fatwas* left and right. . . . Someone that you've never seen—that you don't know, that is nothing but a name—sort of gives you the law of God right there" (Moon, 2002).

This problem came to the limelight in 1998 when a Web site appeared that carried fabricated verses, which were made to look as if they were from the Quran. The verses used derogatory language to insult Islam and Muslims. Bunt (2000) documented the case, saying it was "(p)erhaps the most extreme example of a non-Muslim 'challenge' to Cyber Islamic Environments" (p. 126). The Web site listed a number of fabricated Quranic chapters "*surah*" that were written in classical Arabic and attempted to follow the same meter and rhyme of the Quran. The content though was extremely offensive and humiliating to Muslims. At the time of its inception, the Web site was hosted by AOL (America Online). After several complaints on the part of Muslims, AOL closed the site, saying it was "defamatory in nature" and "particularly targeting Islam." However, the Web site appeared again with a new address and was still on the Internet at the time of reviewing these lines in September 2006. This case, while definitely not the only one, serves to highlight several important issues regarding authenticity and authority on the Internet. With the advantage of allowing every Internet user to become a publisher and voice his/her opinion to the world comes the problem of how to make sure the information you are reading is what it purports to be. My personal advice is to check the contents of the Web site against other Web sites, and if possible, ask a Muslim friend or e-mail a Muslim scholar, or two, or several. And before all else, use your mind and common sense on what seems logical.

Popular Islamic Web sites

We will now move on to an attempt to map out the most popular Islamic Web sites or Web site genres online, and give a few examples or illustrations of their contents. Several criteria were utilized in choosing these sites, including Web searches for the Islamic sites with the top traffic rates, as well as some sites that eminently use the word "Islam" in their names, and therefore might have higher access rates due to their high placement on a search engine when a user searches for the word "Islam." Please note that Web site addresses do change, some Web sites disappear and others get introduced. However, all URLs were correct and functional at the time of going to press.

To begin with, some Web sites provide a list, arranged alphabetically or by topic, of other Islamic Web sites, whereby you are taken to the individual site by clicking on the link provided. An example of such sites is provided by Jannah.org. The word "*jannah*" is Arabic for "paradise." The site provides what it calls a "Mama List"[1] of Islamic Web sites, with over 2,000 listings arranged by alphabetical topics (Figure 6.1). The site allows you to search within the list,

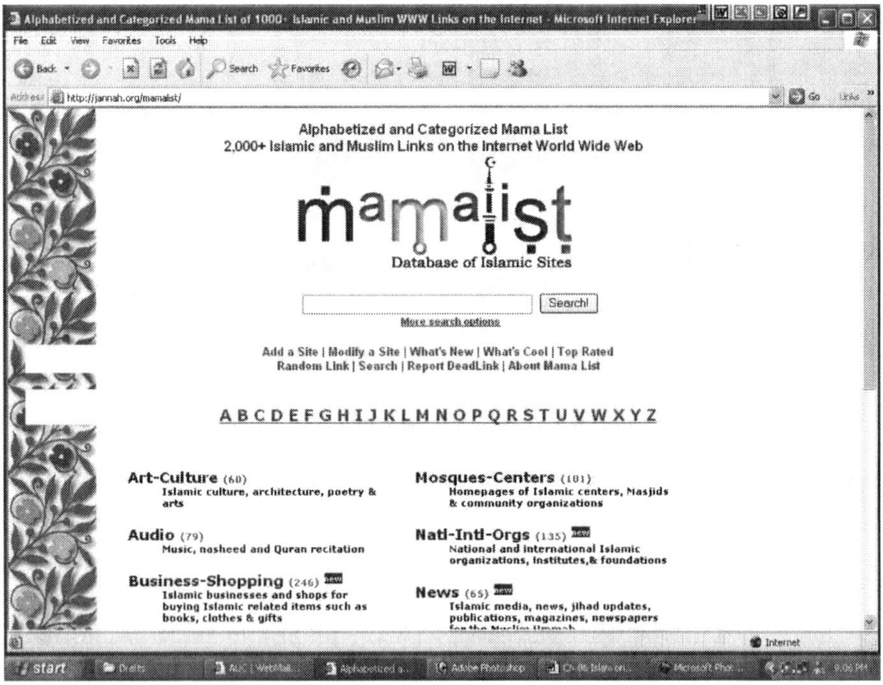

Figure 6.1. The Mama List of Islamic Web sites on Jannah.org.

to add sites, modify sites, or report bad links. Another is Islamic sites.com,[2] which "has been developed with the objective of bringing the authenticated Islamic Web sites together on a single directory." Anyone can fill a brief form to add his/her Web site to the directory, but as the site owners warn, this is not a guarantee for inclusion, as only sites deemed credible by the owners will be added. In their own words, "We thoroughly scrutinize every posted Web site before adding it to our directory." The site also features a section with articles and information about Islam and its tenets.

Islamic Port[3] is another such site offering numerous links arranged by topics. However, the site, which is advertising based, does not list any conditions for inclusion of a Web site address. It states, "Islamic Port is a human edited Islamic Web sites directory, featuring top Islamic Web sites, Hot Islamic Links and serving targeted Islamic Banner advertising by category and keyword." However, the number-one top Islamic Web site listed was a banner advertisement for "Islamic Boutique." Nevertheless, the homepage does list a good number of Islamic links, categorized by topic. The site also carries out a number of polls, apparently to serve its targeted advertising efforts.

Another Islamic Web site list site[4] offers an alphabetical listing of sites. Despite the simple, amateur design, the site still provides links to hundreds of informative Web sites about Islam, including a clear sign that reads, "Dear viewers, please visit islamonline.net for authentic Islamic contents and fatawa."[5] A description of the Islam Online Web site will be provided later in this chapter.

One of the best efforts at organizing Islamic links is a site called Maranao.com,[6] designed by a group of Philippine Muslims working in Saudi Arabia, and copyrighted to the Federation of Maranao Associations in Saudi (FEMAS). The site has an enormous collection of links, arranged alphabetically. It also includes religious news and articles.

Islamic Portals

The next category of Islamic Web sites features comprehensive Islamic portals. Islamic City[7] is one major English-language portal. Highlighting the Quranic verse that says, "Let there be no compulsion in religion" (The Holy Quran, 2:236), the site aims to minimize misconceptions and spread knowledge about the true nature of the religion. The main page is professionally designed, well organized, and includes a huge amount of links. It can take you to a wealth of information on understanding Islam, its tenets and history, Quran and Islamic teachings, worldwide prayer times, U.S. mosque locator, and Islamic

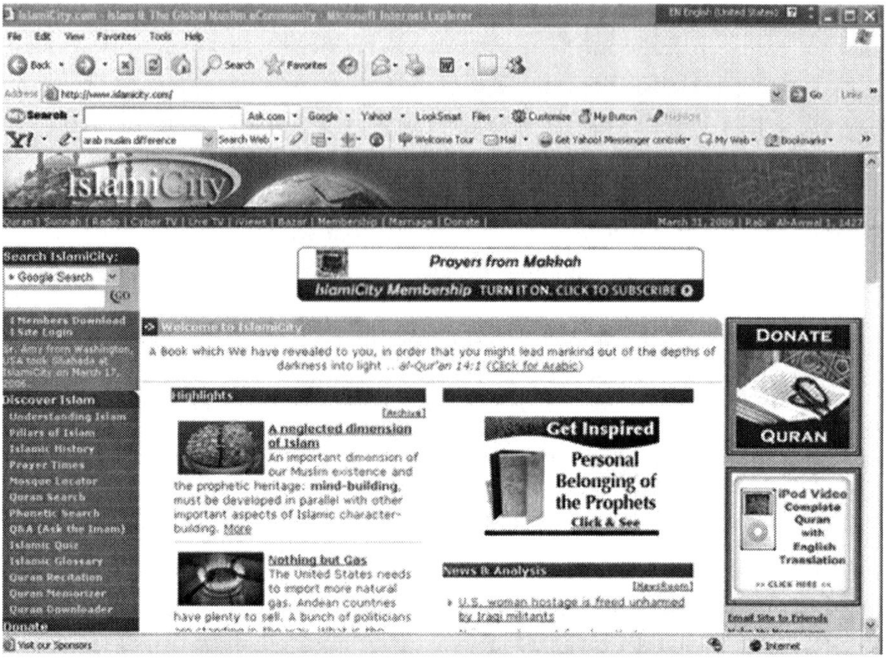

Figure 6.2. IslamiCity.com.

fatwas and questions. It also provides Islamic television programs, live prayers, live radio, a marriage matchmaking service, a bulletin board, screen savers, cooking recipes, as well as opportunities to ask questions to Muslim scholars and to donate or pay the Islamic *Zakat* (charity giving or alms, which is a main tenet of Islam) online by credit card. The portal also offers a section on international and world news (Figure 6.2).

Islamic City reports that it is visited by more than 24 million visitors per year, with over 10 million page views, and 200 million hits. The users come from more than 120 countries, with 58% coming from North America, and 28% from Europe. The site has received several awards and was recognized in a slew of major international media outlets.

Discover Islam[8] is also a portal that introduces Islam in English and French. The homepage of the site is amateurish in design, but the links take you to a number of other Islamic sites that are professionally maintained. This Web site aims at correcting misconceptions about Islam. Besides the regular instructional links, there are also topics such as Islam and terrorism, women's rights in Islam, Islam today, Muslims in the West, misconceptions about Islam, among others.

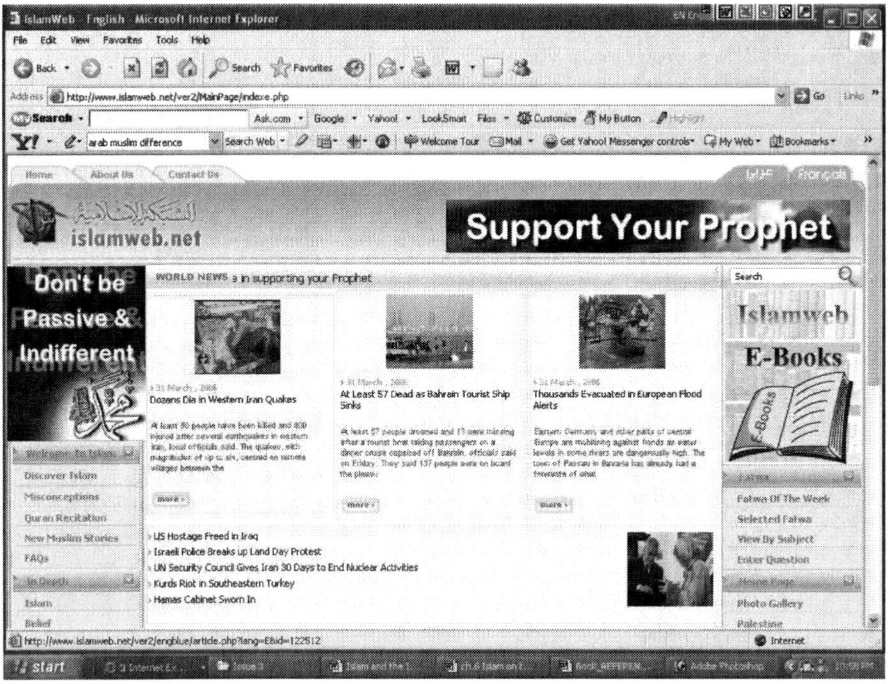

Figure 6.3. Islam Web.net.

Islam Web[9] is another comprehensive portal, offered in Arabic, English, and French. The portal also offers all the primary Islamic knowledge tools, plus sections on family, news, medicine, culture, weather, international media, a kids' corner, a multimedia library, weekly polls, and discussion boards (Figure 6.3).

The Web site of Al Azhar University[10] is another source of comprehensive information on Islam. Al Azhar is the first Islamic university in the world, having started in 988 AD. It was built around Al Azhar Mosque, which was built in 971 AD, and is one of the most influential institutions in terms of preserving the cultural heritage and sciences of Islam. Today, students from all over the Muslim world come to study Arabic language and Islamic laws and traditions at Al Azhar University. Al Azhar Web site offers information about the tenets of Islam, as well as updated research and access to two magazines produced by the University.

The Jannah.org site[11] is another example of a comprehensive Islamic portal. The site provides access to pages for Muslim women, kids, youth, introductory information about Islam for non-Muslims, Quranic recitations, translation, and interpretation. The site also includes an Islamic library, with books, reviews, and articles about the religion, its history, development, tenets, and laws. Web pages about Ramadan, the holy month of Muslim fasting, as well as the pilgrimage (*hajj*) are also featured. A link for a radio show called "Window on Islam" is provided for the live broadcast every Tuesday at 2–3 p.m. EST. The site is designed by a Muslim male, most probably of Arab origin (as perceived from the language and Arabic words used) and apparently living in North America, who designed it while he was a master's student in psychology and computer science. Despite the simple nonprofessional look and design of the Web site, the amount of work that went into building it is quite impressive. As the owner himself describes, it took him years to put it together. He also tackles the issue of credibility of information and offers a few wise words of caution to anyone looking for information on Islam. In his own words:

> Over the course of 7 years and [sic] developing this Web site I have some observations for Muslims and those seeking information about Islam: First, BE CAREFUL about what you read. I've been through hundreds of sites and it is amazing that some will say "TRUE ISLAM" or even have nice Islamic titles, and they turn out to be wack information pages [many written by people against Islam]. You want to read "the other side," you say. Why read a page on Islam by a non-Muslim? Most contain inaccuracies, misconceptions and strategically present Islam in a bad light on purpose without giving the complete explanation and understanding of issues. Learn about Islam with an open mind from the sources of Quran and Sunnah and practicing Muslims and decide FOR YOURSELF what you think about it. The nature of the internet is such that anyone can publically [sic] advance their views freely [a good thing] and claim to be anything, but make sure what you are reading is from authentic Muslim sources. The internet should only be used as a resource and anything suspect should be checked out and explored; better sources being your local Imam, knowledgable [sic] Muslims, Mosque, Islamic center or Muslim Students Association.

Some portals are popular in the Muslim world for the popularity of the Islamic sheikh or scholar behind them. One such example, perhaps the most popular site in the Muslim world, is Islam Online,[12] which is supported and maintained by a large group of Muslim scholars headed by Sheikh Yusuf Al Qaradawi, one of the top authorities in the Muslim world today. The portal, offered in both Arabic and English, is very comprehensive and is perhaps the most sensitive to the world topics of the day. The site is run by a staff of over 200 people from different backgrounds. In addition to all the usual Islamic information, the site also features large sections on science and medicine, psychology and cyber counseling, political news and current affairs, women's rights, parenting, the Internet and IT technologies, arts and culture, in addition to live *fatwas* and live discussions. Each of these sections is handled by professionals, a lot of them PhD holders, in their respective fields. At the time of writing the first draft for this chapter in March 2006, the site featured sections on the Danish cartoons crisis,[13] European Muslims, Avian flu, Iraq, the Israeli elections, the movie "Syriana," water preservation, among others. According to the site's "About Us" page, it targets Muslims and non-Muslims alike, and aims to "strive to provide you with all the information you need about Islam and its civilizations, the universe and its changes, current affairs and their analyses, and general information and services that one cannot do without in the 21st century." (Figure 6.4).

The site played an important role in the aftermath of 9/11, when it increased its basic English language content to include "even the most basic tenets of the religion," according to its business development manager. She claims that the site witnessed a huge increase in visitor numbers during this period, with page views increasing from an average of 24 million to 150

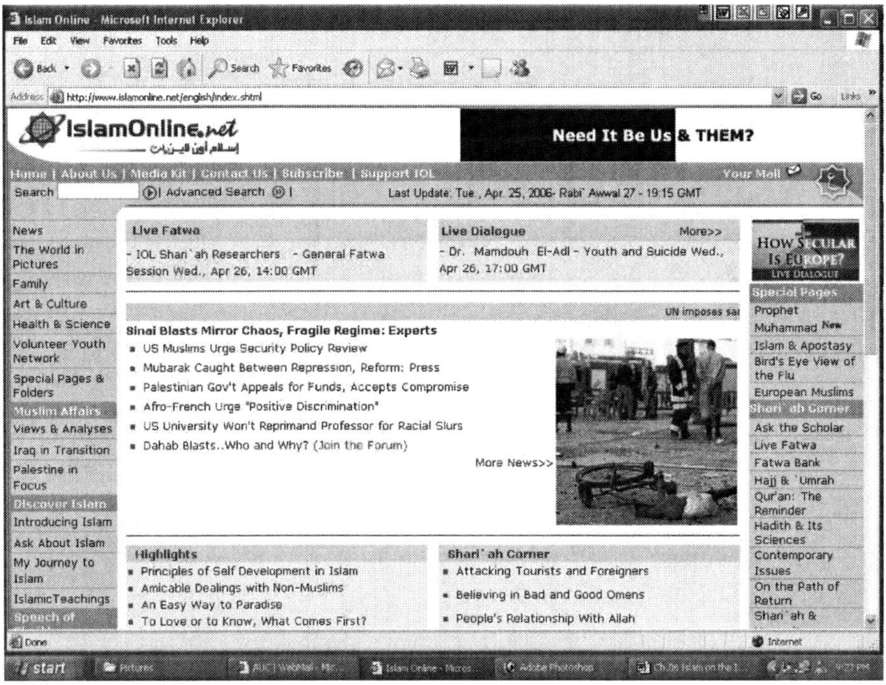

Figure 6.4. Islam Online.net.

million per year (El-Kashef, 2005). Islam Online stresses on the values of Islam as a religion of moderation. For example at the time of writing, it stressed on the *fatwas* that while peaceful demonstrations against the Danish cartoons are permissible, violent acts are extremely *haram* or forbidden in Islam.

Another influential Muslim figure with a popular Web site is Amr Khaled.[14] Khaled is a 38-year-old Muslim activist, with frequent appearances in the media and several collections of audio and video Islamic material in his name on the market. Khaled is not an Islamic scholar though, having graduated from Cairo University's Faculty of Commerce. As such, his vast popularity among young Muslims has caused a lot of controversy over the credibility of his Islamic talks and lectures. In recent years, partly in response to the criticisms, Khaled has tried to stay away from Islamic preaching per se, dedicating his talks more to morality, lessons learned from the life of Prophet Mohammad, and national development in the name of Islam. His Web site is available in 16 languages, including English, French, Spanish, German, Italian, Danish, Farsi, Hebrew, Chinese, and Japanese.

Most recently, in reaction to the Danish cartoons crisis, Khaled gathered a delegation of 42 Muslim scholars and 25 Muslim youths and went to Denmark for what he called the Copenhagen Conference. The conference was hosted by the Danish Youth Council, involving the Danish Institute for International Dialogue, as well as universities and clubs reaching out to more than 800,000 members, which Khaled said were mostly engaged through Internet communications. The Muslim delegation held talks, debates and discussions about Prophet Mohammad, his life, and teachings. As a result, a youth exchange program has already started between Muslim and Danish youths, online communication is ongoing, and several other projects are in progress.

Some Islamic Web sites are designed and maintained by Westerners with an interest in Islam. An excellent example is the Web site of Professor Alan Godlas[15] of the University of Georgia's Department of Religion. Professor Godlas has designed his Web site "to be of use for non-Muslim and Muslim students and teachers at all levels as well for members of the general public who wish to get a nonpolemical view of Islam (including Sunni Islam, Shi'ism, and Sufism) and to a lesser extent of Judaism and Christianity." Highlighted in an article in *The Chronicle of Higher Education* (Read, 2001), the Web site is a comprehensive educational tool about Islam, including not only the regular links to Quranic recitations and translations, but also links to Web sites about women's rights, Islamic art, architecture, music, the mystique of Sufism, as well as academic and scholarly articles about Islam and scientific knowledge and

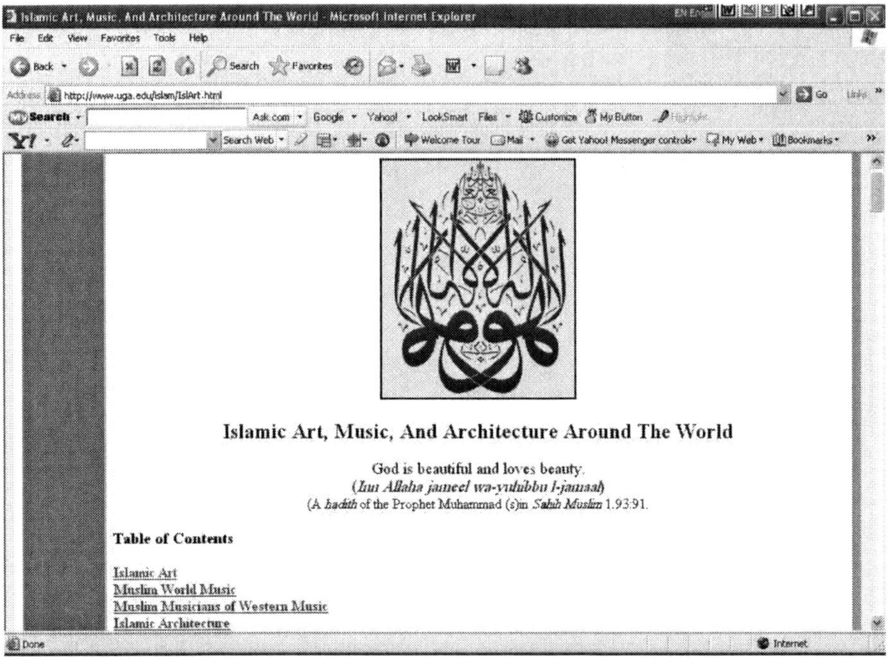

Figure 6.5. A Web page on Islamic art and architecture off the Godlas Web site.

research, and Islamic habits and statistics from each country of the Muslim world. The site is characterized by a clear scholarly taste, whereby articles chosen are not mere opinions of a Mohamed Doe but rather an academic piece of writing, supported by objective arguments (Figure 6.5).

Like many other Islamic sites, Professor Godlas's Web site took a design turn after the events of September 11, 2001, whereby the professor added a multitude of links about such topics as Islam and violence, Islam and Jihad, Islam and peace, Muslims in Iraq, and human rights in Islam among others. His page about Jihad and violence in Islam[16] starts with a quotation by renowned Princeton Professor Bernard Lewis, who is famous for his anti-Islamic views, saying, "At no point do the basic texts of Islam enjoin terrorism and murder. At no point do they even consider the random slaughter of uninvolved bystanders—Bernard Lewis in 'License to Kill' (1998)." The site notes the common mistaken translation of Jihad in Islam as "holy war" and offers multiple explanations to the true meaning of the concept based on Muslim and non-Muslim scholarly work. Another page off the Web site is dedicated to 9/11, Bin Laden, and Islam.[17] The page offers 37 articles on the topic, that cover aspects such as

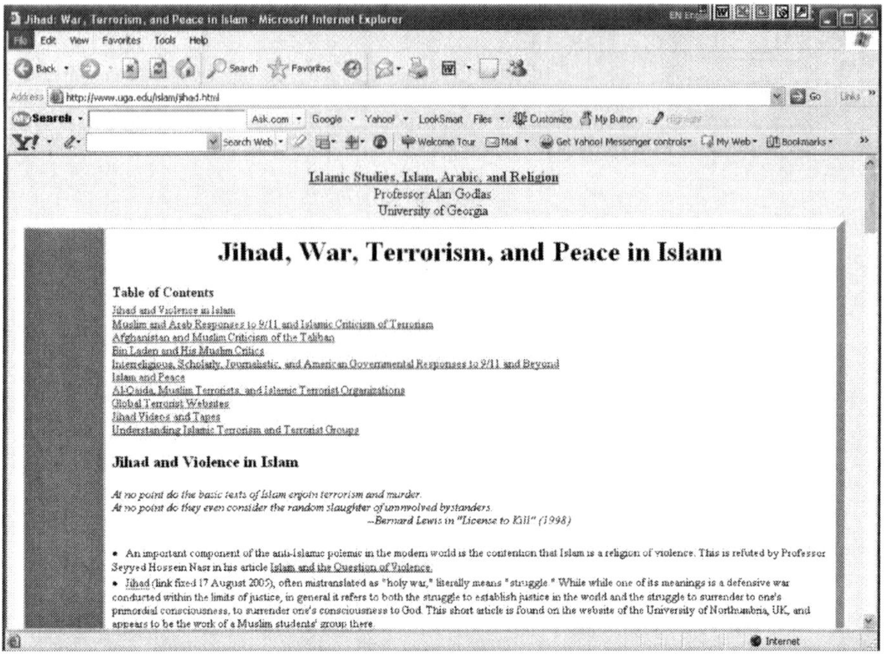

Figure 6.6. The Godlas Web site features important topics about Islam after 9/11.

Muslims' and Arabs' reactions to 9/11, Muslim criticisms of Bin Laden, Afghani and Muslim reactions to the Taliban, and scholarly analysis of Islam and 9/11. Indeed, the material presented is so well chosen that *The Chronicle of Higher Education* noted the popularity of the Web site following 9/11. Read (2001) reported on Professor Godlas's Web site statistics, saying that the number of visitors to the Web site has increased from an average of 250 a day before 9/11 to several thousand visitors per day after the attacks. The credibility and visibility of the Web site were enhanced by a link to it posted on the Web site of National Geographic (Figure 6.6).

Another interesting aspect of the Web site is the inclusion of a wealth of information about Muslim rock singers, including Cat Stevens (now known as Yusuf Islam),[18] Richard Thompson, and Everlast. The main purpose of his Web site, says Godlas in the *Chronicle* article, is that "it's crucial for people to understand the difference between militant extremism in the name of Islam and Islam per se. And to the degree that people don't understand that [distinction], Muslim-Americans in particular will receive the brunt of any prejudice that derives from that misunderstanding."

Benefits to Muslims

Perhaps an important function that the Internet has afforded to Muslims worldwide is the ability to connect, communicate, and discuss issues on an unprecedented scale, and without worrying about obstacles of time and geographical boundaries. Discussion boards, blogs, and simple Internet applications now available on most Islamic portals allow not only conversations but also posting questions of Islamic law to Muslim scholars.

Nevertheless, it is this author's opinion that the Internet is bound to be of a more direct religious benefit to Muslims living in Western countries, much more so than Muslims living in Arab or Muslim countries, particularly with the obvious challenges posed by the digital divide. There are several factors supporting this opinion. Muslims are a small minority in the United States and a lot of European and Western countries. Connecting online with people who share a common value with you, especially one as revered as religion is to Muslims, is much more meaningful where there aren't many of these people around in "real" life. When you are a minority, by definition, access to people who share your core beliefs and who are more likely to understand things about you that others won't, becomes much more valuable. For a female Muslim in a Western community, the majority of people around her might not necessarily understand, or appreciate, or tolerate a particular clothing pattern that she follows, if she happens to wear a head cover (veil or *hijab*) or a long conservative dress (*galabiya*). Western friends might not understand why a Muslim might refuse to share an alcoholic drink or a meal of pork with them. Even after explanations by the Muslim person, he/she still remains a "stranger," someone who acts differently from the majority of the rest. It is much harder for that person to become "one of the gang." It is in these instances when connecting with people who share your values would mean the most to you.

Arabs living in Western countries have another reason to benefit from online communications with other Arabs and Muslims: the ability to "speak" in Arabic. If you happen to live in a place where no one understands or uses your mother tongue, you tend to miss it. Communicating online in Arabic over e-mail, discussion boards, and chat applications (both text and audio) provides a valuable link to "home," in terms of language, values, culture, and everything else.

Another reason why the Internet would be more valuable as a tool for dialogue for Arabs and Muslims living in Western countries is that it is much easier to have face-to-face access to an Islamic scholar, a sheikh or an Imam, in

Muslim countries than it is in Western countries. There are a good number of mosques in every suburb in Muslim countries. Cairo, for example, is known as the City of a thousand Minarets.[19] With mosques abundant and conveniently located, many Muslims choose to pray frequently in a mosque rather than at home.[20] After prayer, Muslims have a chance to talk with the sheikh or Imam in the mosque, and ask him any questions they might have regarding Islam or Islamic laws. This same opportunity, with that much convenience, is not available to Muslims living in the West, who sometimes have to drive for miles to reach a mosque, and therefore might do it only on major religious occasions. On such days, the mosques are usually very crowded and there is little chance of talking one-on-one with the Imam. And so the Internet presents a wonderful alternative, allowing them to ask their questions not necessarily to the Imam who happens to be at the mosque but to a variety of Muslim scholars on the multitude of Islamic Web sites online.

Muslim religious occasions are another time when the Internet is valuable to Muslims living in the West. Muslim holidays and special occasions often go unnoticed in the West. Much unlike Christmas, Easter, Thanksgiving, Rosh Hashanah, Yom Kippur, or Passover, very few people (particularly before 9/11) have heard of Ramadan,[21] Eid al-Fitr,[22] or Eid al-Adha,[23] or know when these major Muslim occasions are observed. Life becomes very lonely if you are living in a society where nobody has heard of the occasions that mean a lot to you. Imagine yourself living in a country where no one says "Merry Christmas" or "happy Thanksgiving" or "happy Hanukkah" to you as you celebrate these major holidays solemnly on your own. Not a very nice celebration! That is how Muslims in the West feel on their religious holidays, and, once again, the Internet now comes to the rescue. E-cards or an e-mail from a friend back home makes a big difference on such occasions.

Of course, the Internet plays another role for Arabs and Muslims in the West by connecting them to other Arabs and Muslims. Discussion boards and chat rooms help Muslims get to know others in their own area, whom they could eventually meet face-to-face. Most of the Muslim portals include discussion boards and chat functions, and some even provide matchmaking services, where Muslims can get to know others, particularly those living in close proximity.

Before I end this chapter, I want to remind you that acquiring an accurate picture of Islam through the Internet requires a bit of research. Do not be satisfied with the first Web site you browse through, even if it is a major portal. *Fatwas* and religious questions discussed on any one portal usually come from a variety of Muslim scholars, and some questions, indeed many questions, would be answered differently by a different scholar. Credibility of the information presented is key, and if something does not make sense to you, check it out on a different Web site.

It is the core nature of Islam as a religion to be based on a direct relationship between an individual and God, with no interference from religious authorities or Muslim scholars of any kind. This is one main difference between Islam and the other monotheistic religions: the religious clergy can and should provide advice and answers, but their word is just that: advice and answers. Their word is not a command and is not obligatory. A Muslim is only bound by the Holy Quran and the official *hadith*[24] and teachings of Prophet Mohammad, the interpretations of which differ, sometimes in a major way, among scholars and individuals. The Prophet has advised Muslims to, when in doubt, follow their hearts. So again, credibility of information is key, particularly on the Internet. If something does not make sense, check other sources. Islam is a peaceful, moderate, and sensible religion. So make sure to think things that you read on the Internet over, let your common sense and your mind be the judge, and follow your heart.

· 7 ·

TO E- (FREELY) OR NOT TO E-: ARAB GOVERNMENTS AND THE ONLINE FREE FLOW OF INFORMATION

Freedom of expression has always been a thorny issue in the Arab world. So much so that some have argued that the relatively late and slow launch of the Internet in some Arab countries is primarily due to a desire to control information on the part of the governments of these countries. Franda (2002) reports that in 1997–1998, Iraq made it illegal for citizens to access the Internet from within its borders, and that until June 2000, Syria did not allow its citizens to access local ISPs. In Saudi Arabia, before public access to the Internet was available in 1999, citizens who wanted to access the Net had to log on through an international phone call to an ISP in Bahrain or another neighboring country. According to Human Rights Watch (1999), more than 30,000 Saudis did access the Internet regularly through international phone calls to ISPs. Human Rights Watch also reported that the Kingdom has had access to the Internet since 1994, but barred its citizens from public access until the government ensured a good system for controlling what it referred to as "the bad things on the Internet."

Arab countries have large variations within their political systems and the margin of freedom that is allowed in each country. Freedom of expression on the Internet is no exception. The margin of online free flow of information varies from relatively liberal in Jordan, Morocco, and, until recently, in Egypt, to strict in Saudi Arabia, the United Arab Emirates, and Tunisia.

In Saudi Arabia, Internet access goes through King Abdul-Aziz City for Science and Technology (KACST), which services universities and private ISPs that the country started licensing in 1999. Those ISPs operate as intermediaries between consumers and the Saudi Telecommunications Company, which has monopoly over telecommunications services in Saudi Arabia. Since all Internet traffic goes through KACST, it is easier to control the medium. Saudi authorities were proud to have censored 200,000 Web sites in the first 18 months after the service was opened to the public (Al Khuwailadi, 2001; Open Net Initiative, 2004). Human Rights Watch (2005) reports that most of the sites blocked by the Saudi government are pornographic in nature or contained materials critical of Saudi politics, religion, or Gulf regimes. Saudi officials themselves admit to overcensoring sometimes, particularly when it comes to medical sites since they use censorship software that depends on blocking sites with certain keywords. Consider, for example, if someone were to look for information on breast or prostate cancer, he/she would not be able to access any sites, because they would be mistakenly blocked by censorship software as pornography sites.

The Saudi government never tried to hide its intentions to censor the Internet. Rather, they take pride in "protecting" the Saudi citizens. Before 1999, Saudi officials were quoted frequently regarding the technicalities of what they viewed as "the region's most ambitious plan to block the flow of undesirable data online." The five-year delay between having Internet access in the country and having it available to the public was usually attributed to working on a system that would protect the Saudi citizen from "inappropriate" material on the Internet (HRW, 1999, 2005). Among the possibilities discussed was the idea of having Web sites "white-listed" rather than "blacklisted." In other words, the Saudi government would approve a list of sites, and those would be the only Web sites the user is allowed into. All other sites are blocked by default until a request is sent to the government to review the site, and if approved, add it to the white-listed sites. Fortunately, this scenario was never put into action. There is now talk that Prince Walid Ibn Tallal is planning to introduce an alternative Arab, precensored World Wide Web to serve Arab audiences. At any rate, strong censorship on Internet content is still practiced in Saudi Arabia today, whereby citizens are prohibited from accessing numerous Web sites, including "carrying out any activities violating the social, cultural, political, media, economic, and religious values of the Kingdom of Saudi Arabia" (Decision # 163, May 1998). It has been reported however that several Saudi ISPs are investing in the services of professional hackers to allow them to circumvent censorship and offer access to blocked sites to their users anonymously. Other ISPs have opted to connect to the Internet through a foreign route to avoid government censorship of Web content (HRW, 1999; Privacy International & the GreenNet Educational Trust, 2003).

The Web site of the Internet Services Unit at KACST includes a detailed description of the "filtering" process, how it is done, what types of sites are blocked, as well as request forms for blocking and unblocking sites that Saudi citizens can fill out. The site claims that it receives hundreds of blocking requests every day. Saudi Arabia uses a U.S.-made commercial software called Secure Computing's SmartFilter. One study (Open Net Initiative, 2004) tried to assess what type of Web sites are blocked by the Saudi government. After attempting to access 60,000 sites in 2002, 2003, and 2004, the researchers concluded that although the government focuses on sexually explicit sites, as well as sites for gambling, homosexuality, religious conversions, and filtering circumvention tools, other types of sites are also blocked. They found considerable blocking for sites regarding political opposition groups, the Bahai faith, the Holocaust, free Web hosting, women's rights, and any sites that include images of "provocative" attire. The researchers did report examples of erroneous overblocking, for example, a Hawaiian church site was mistaken for a pornographic site, and a women's human rights site was blocked because it had a picture revealing a part of a woman's body to show marks of torture. Many people reported not being able to access articles in medical journals, because they contain words such as "sex," or "genitals," or "breast." Actually, if you think about it, most, if not all, medical sites will include either a picture of a human body or words that describe human organs. Potential for overblocking is also enormous when it comes to any type of research online, because the research report or instrument is more than likely to include the word "sex," albeit it meaning "male/female."

The Saudi government often tries to show the public's support for its censorship policies regarding the Internet. The KACST Web site reports on an "old" undated user survey that shows that while 45% of the Saudi Internet user population thought there is "too much" blocking of sites, 41% believed it was "reasonable" and 14% believed it was "not enough." In the "new" user survey, conducted in 2002, only 16% of respondents believed blocking was a problem. Saudi officials have also often reported wide public support for online censorship. In 2001, they reported that the KACST received an average of 500 block requests and 100 unblock requests a day. They said that 30% of the block requests are carried out, as opposed to only 3% of the unblock requests. That is, the Saudi government blocked 150 sites daily, and unblocked three (Al Khuwailadi, 2001).

While Saudi Arabia is the most restrictive Arab country in terms of Internet use policies and censorship, Syria was also very restrictive regarding the Network until the passing away of President Hafez Al Assad in June 2000. As a matter of fact, it was only in June 2000 that Syria allowed public access to the Internet. Before that, Internet connectivity was illegal in the country although it was available since 1997. However, Syrians still found their way into the network, mainly through foreign ISPs. One Syrian citizen was quoted in *Newsweek* in 1999 to say, "Internet connectivity is not allowed (by law), but we're all connected." Franda (2002) reported that the state-owned Syrian Telecommunications Establishment (STE) had started accepting applications for e-mail accounts in 1999 when it was still illegal, because the government realized that thousands of Syrian citizens do access the service through ISPs outside the Syrian borders, particularly in Lebanon and Jordan. Before legally permitting Internet access in 2000, the number of users in the country was estimated at about 30,000 out of a population of about 16 million (UNDP, 2002).

When current Syrian President Bashar Al Assad took over in 2000, many hoped he would open up the world of IT to the people. While the country was still under his father's rule, Bashar Al Assad served as Chair of the Syrian Computer Society, and had often publicly advocated public access to the Internet. He has declared several times that the computer has to become an integral part of the Syrian and Arab culture, and that computer illiteracy is just as bad as regular illiteracy. Indeed, in the first two years since he came to power, Internet users in Syria increased drastically from a mere 30,000 in 2000 to 226,000 users in 2002 (UNDP, 2002, 2004). However, in terms of Internet control, Human Rights Watch (2005) reports that Syrian authorities censor information freely on the Internet. According to the HRW report, "the government tampers with the very fabric of the Internet, restricting the use of the basic electronic protocols that allow people to send e-mails and construct Web sites" (p. 1). With emergency laws having been declared in Syria for over 40 years, the government is able to depend on a set of oppressive laws and measures to sacrifice personal civic freedoms in the name of national security. The government therefore heavily censors the online flow of information, and monitors and censors e-mail. Efforts of Web site blockage are focused on Arabic language sites.

The Syrian government has detained and sometimes tortured people for their online communications, whether through e-mail, blogs, or on the Web, and sometimes even for forwarding political jokes. The report details five such cases, where Internet users spent time in jail. Their sentences ranged from nine months for forwarding a political joke that the government deemed disrespectful of the system, to three years for forwarding news or writing blogs of the same nature. These victims of free speech were accused of "publishing lies," writing articles "that harmed the image and security of Syria," using the Internet to deliver "false news outside of Syria," or even "writing that threatens the security of Syria and her relations with foreign states." As of September 2006, three cyber dissidents were imprisoned in Syria (HRW, 2005; Reporters Sans Frontieres, 2004; RSF, 2006).

Syrian Internet activists told Human Rights Watch that security officers monitor Internet cafes by having officers roam near them in plain clothes, watching the Internet user traffic in terms of who goes in or out, and snooping over people's shoulders to see what they were doing on the Internet (HRW, 2005).

Despite these restrictions, the report says that the Internet is still a healthy outlet for freedom of expression in Syria. In a country where all media are government-owned and heavily controlled, Syrians take to the Internet to attempt to "talk" freely behind the government's back, or access information the government would not want them to know about. One way that they resort to is to access the Internet through ISPs in (or leased lines to) neighboring Lebanon, where they can get a censorship-free Internet. Others try to post their views or blogs using software that make sure they stay anonymous.[1] Trial versions of these software are often available on the Web, making it accessible to a bigger portion of the Syrian Internet users. In other cases, users attempt to make sure they stay one step ahead of the government. For example, one blogger who got his Web site blocked kept sending a daily bulletin by e-mail to more than 16,000 addresses. When the Syrian authorities realized what he had done, they blocked his e-mail address. He changed it. They blocked it again. He changed it again. They kept playing this cat and mouse game for 40 days, each day with a new e-mail address, until the government finally gave up on him (HRW, 2005).

Syrians are still happy with the margin of freedom that the Internet offers them, and hope they can use it to instill more democracy in the society. They are hopeful that change will come, either through the government giving up on Internet censorship or through a change in regimes. As one Syrian human rights activist told Human Rights Watch, "The Internet is the only way for intellectuals to meet and share ideas in Syria today" (HRW, 2005).

Tunisia probably has the most detailed legislation for the Internet in the Arab world. The country modified its press law to account for the new medium, and issued a Telecommunications Decree and an Internet Decree in 1997. Despite efforts from the government to enhance Internet facilities and penetration rates in the country, these regulations have imposed much restraint on freedom of information flow in Tunisia. Internet regulation in Tunisia requires every potential ISP to apply for a license from the Ministry of Communications before it starts operating. However, to review the application, a committee is formed that includes members from the Ministry of Defense, Ministry of Interior, as well as telecommunications, IT, and computer science experts. Once in operation, every ISP has to specify a director, who "assumes responsibility . . . for the content of pages and Web pages and sites that the ISP is requested to host on its servers." Furthermore, ISPs are required to submit a monthly report to the government, including a list of subscribers. The director of the ISP is also responsible for maintaining "constant oversight" on the online content that the ISP hosts, to make sure information on the servers is free of anything that endangers "public order and good morals" (HRW, 1999, 2005).

All Internet flow in Tunisia goes through the state-owned agency, the Tunisian Internet Agency (Agence Tunisienne d'Internet—ATI), which functions under the auspices of the Tunisian Ministry of Communications Technology. It provides and controls the backbone of the Internet in Tunisia. If an organization or an educational institution applies for Internet access from ATI, the same regulations apply to it. Furthermore, institutional users have to sign a contract, which specifies that they will "use the Internet only for scientific, technological or commercial purposes that are strictly related to the activity of the client, in strict conformity with the rules in effect." The ATI has the right to conduct "site visits" to check that Internet use in the organization is in accordance with the contract. This basically means that such users do not have the right to send a personal e-mail using their Internet connection, because it would violate their contractual agreement with ATI and make them liable to potential legal action against them, including fines and/or jail sentences (HRW, 1999, 2005).

Because Internet flow goes through a central route (ATI), it is easy to monitor such content, a benefit which the Tunisian government takes advantage of. The government states that it monitors content to protect citizens from "inappropriate material," particularly pornography. However, some reports indicate that other sites do get blocked, including sites that have critical information about Tunisia or the Tunisian government as well as some human rights sites such as the Tunisian League for the Defense of Human Rights and Reporters Sans Frontières (HRW, 2005; RSF, 2004). In 2002, the government increased online monitoring, including tracking down and arresting users who repeatedly try to access blocked Web sites or post information of a critical nature. Perhaps the case that got the most publicity was the jailing of journalist Zouhair Yahyaoui, who was sentenced to 28 months in jail in June 2002. Yahyaoui was an Internet activist who, under the pen-name "El Tounisi" (Arabic for "The Tunistian"), published *TUNeZINE*, an online magazine that was increasingly critical of the government. After various outcries from human rights organizations around the world, he was released in November 2003. He later died of a heart attack at the age of 36 (HRW, 2005; RSF, 2004).

Other documented cases of cyber detention in Tunisia include Mohamed Abou, a founding member of two Tunisian human rights organizations. Abou had published an article on the Internet one night before his arrest criticizing the Tunisian president and accusing him of corruption. According to Human Rights Watch (2005), "As if to disguise its persecution of Abou for his lèse majesté, the government prosecuted Abou on dubious charges stemming from his alleged assault of a female lawyer in 2002 and for publishing an article six months earlier critical of prison conditions in Tunisia." He was later charged with "publishing false news capable of disturbing the public order" (p. 145). Abou is currently serving a three-year prison sentence.

Despite the tight Internet regulations and the consequences they bring, Tunisia was praised by ICANN, the worldwide Web-regulating organization, for its efforts to spread the Internet (as was illustrated in chapter 4). The country was chosen to host the second phase of the World Summit on the Information Society (WSIS) in November 2005, a choice which some Internet and freedom of expression activists condemned given the way the Tunisian government has been dealing with the online free flow of information (HRW, 2005; RSF, 2004).

We now shift our attention to the United Arab Emirates, the country with the highest percentage of population connectivity in the Arab world. Although Internet users are a mere 1.4 million, they account for about 36.5% of the population. However, according to Human Rights Watch and Reporters Sans Frontiers, the United Arab Emirates is also a leader in advocating censorship on the Internet through filtering software. The country's sole Internet service provider is Emirates Internet and Multimedia (EIM), which is part of the Emirates Telecommunications Corporation, also known as *Etisalat*. Emirati citizens dial up to EIM's proxy servers in order to access the Internet. The servers have lists of banned Web sites. If users request access to one of these sites, they are denied access. Similarly, users are denied access if they type in a URL of a site which is determined by the proxy servers to have "inappropriate" materials (HRW, 1999; RSF, 2004) (Figure 7.1).

Government officials of the United Arab Emirates admitted to Human Rights Watch that they do administer the filtering system, but said that censorship was only used against pornographic sites, and not against political or religious sites. However, Reporters Sans Frontiers reported that, in actuality, various types of online publications, religious sites, and dating sites are also blocked. Like in Saudi Arabia, medical and scientific sites that have pictures of naked body parts are also censored. The official who spoke to Human Rights Watch admitted though that users can still access pornographic sites, or any other sites, through a foreign-based ISP, or if they access the Internet via a private leased line, which is the case in most work places in the United Arab Emirates. Such leased lines do not go through the proxy servers, and can therefore circumvent the censorship system imposed by the government (HRW, 1999; RSF, 2004).

The United Arab Emirates however is one of the Arab countries that believe in the power of the Internet for economic development. Most if not all ministries in the government maintain sophisticated Web sites in Arabic and

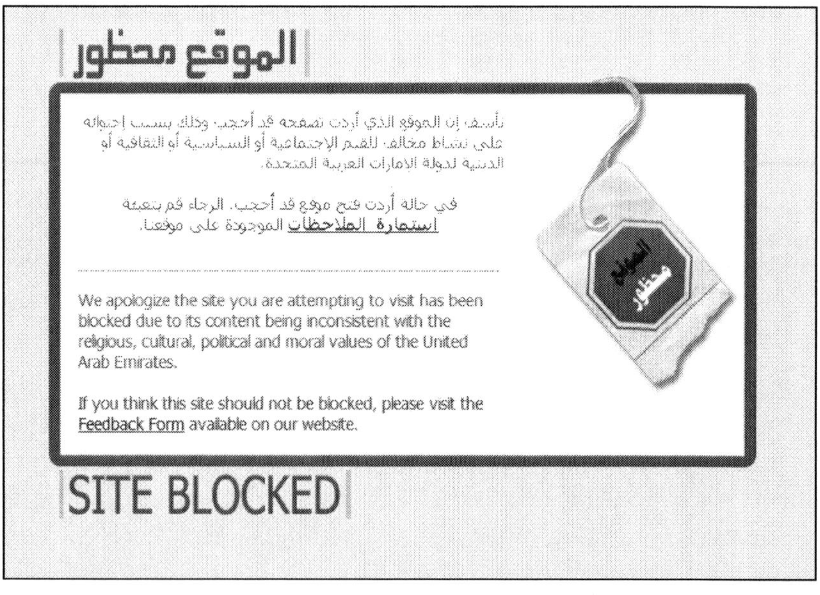

Figure 7.1. A blockage message as it appears from the United Arab Emirates.

in English. Corporate and commercial Web sites are also numerous and of a high quality. The government does require a license for commercial companies to launch a Web site, but rarely have any such requests been denied. Officials say licensing is required to prevent commercial fraud and protect against copyright infringements. As was discussed earlier, the United Arab Emirates has also established an Internet city, one of only two in the Arab world, to attract local and foreign investments in the field of IT.

Some Arab countries have shown more flexibility in their dealing with the Internet. Foremost among these countries are Jordan, Morocco, and, until recently, Egypt. In Jordan, since the Internet became available to the public in 1996, the Internet has enjoyed strong backing from the Kingdom's government headed by King Hussein at the time and now King Abdullah II. The Internet enjoys a good deal of freedom in Jordan, with no government approval or licensing required for access or Web site development. However, all ISPs in Jordan pass through the central government router, which makes it easy for the government to monitor the online flow of information if it so desires. After September 11, 2001, there have been reports that Jordan started monitoring e-mail communications and some Web content, but no reports of censorship or blocking were made. Jordanian citizens reported that on several occasions, they used the Internet to gain access to news and stories that were either blocked or were not receiving enough coverage on more traditional media outlets. In a disturbing move however, the Jordanian government pressed charges on January 5, 2006, against Jamil Abu Bakr, the editor-in-chief of an opposition party Web site, for posting two articles by two Parliamentary members of his Islamic Action Front, on the party's Web site. Although the articles were posted more than a year earlier, Abu Bakr was charged with "belittling the dignity of the Jordanian state." The move has raised concerns that Jordan might have regressed to a state of more government control of the Internet (HRW, 1999, 2005, 2006).

In Egypt, the government does not interfere in the flow of online information. Although the country has been operating under emergency laws since the assassination of the late President Anwar El Sadat in 1981, the status of press freedom has seen improvement in the past years. With regard to the Internet, the government has not interfered to block access to any Web sites, including Web sites of human rights organizations worldwide or Web sites of any political inclination,[2] including some that harbor harsh criticisms of the government and the president on their Web sites. Human Rights Watch (2005) called the Egyptian efforts to spread Internet access "impressive," and reported that the government does not practice online censorship. However, the organization still draws attention to several incidents. The first case, which brought freedom of expression on the Internet to the limelight, occurred when police arrested 55 allegedly gay men, at a party on the Queen Boat on the Nile in June 2002, and accused them of habitual debauchery. The Internet has been used by the group to publicize their activities on a Web site they developed, which could have been how the police found out about the time and place of the party. The 21 men who were convicted with one-to-two-year sentences were later released on bail after President Hosny Mubarak rescinded the verdict. Although they were not convicted because of any Internet-related activity, Internet activists still worried about the freedom of the Internet information flow.

Two other cases warranted activists' attention. In April 2003, authorities raided the house of a political activist, Ashraf Ibrahim, and confiscated his computer. Two days later, after turning himself in, he was detained under Egypt's emergency laws, and eventually charged with "harming Egypt's reputation by spreading abroad false information" about human rights in Egypt (HRW, 2005, p. 29). He was also charged with belonging to an illegal organization called the Revolutionary Socialists. Ibrahim was acquitted by the court in March 2004 and freed immediately thereafter (HRW, 2005; RSF, 2004).

More recently, in October 2005, a 21-year-old law student was detained after posting a highly offensive piece on Islam and Muslims on his blog. Abdel Karim Nabil Soliman, also known as Karim Amer, is a women's rights activist blogger, who has a reputation for using derogatory language. No charges were pressed against Amer, and he was released 18 days later. It was interesting to note that although many reported being highly offended by Amer's language on the blog that allegedly resulted in his arrest, those very same people denounced his detention as a prisoner of electronic opinion and strongly called for his immediate release. Amer's blogs, including the one in question, are still available on the Web (and were at all times), and he still writes regularly on his Web site and on others (Abdulla, 2006).

The year 2006 witnessed more obvious cracks on blogger freedoms in Egypt to the dismay of free expression advocates. All in all, six bloggers, including one female, were arrested while participating in peaceful demonstrations in solidarity with the Egyptian judiciary. Most notably, award-winning blogger and free speech activist Alaa Seif al-Islam Abd el-Fattah was arrested on May 7. He remained in jail for 45 days before being released. Alaa is a pioneer blogger, who maintains a bi-lingual, outspoken Web site with his wife blogger Manal Hassan (manalaa.net).

On April 26, blogger Malek Mostafa, better known as Malcolm X was also detained during another peaceful protest in solidarity with the Egyptian judiciary, and remained in prison for 39 days. Similar scenarios occurred with Asmaa Ali, who was detained for 28 days; Mohamed Sharkawy and Karim el-Shaer, who were detained for 56 days; and Mohamed Adel, who was detained for 31 days (RSF, 2006).

While the blogs maintained by these activists were not censored by the government or brought into official police investigations, the incidents alarmed blogger communites and free speech activists worldwide. Since these bloggers were detained without charges, it seems that the government is cracking down on outspoken online activists through offline incidents. However, at the time of reviewing these lines in September 2006, no more bloggers were detained and those freed continue to be as outspoken as ever on their respective Web sites.

Still, Reporters Sans Frontiers and Human Rights Watch have testified on several occasions that freedom of expression in Egypt seems to be healthier online than offline (HRW, 1999, 2005; RSF, 2004). Internet users can remain anonymous if they wish to, and there are no licenses required to use the Internet or to develop a Web site. There is no censorship of the Internet through the many Internet cafes throughout the country or through any of Egypt's over 300 ISPs, who can access the Internet via the Egyptian government's hub or any other international gateway they choose. This technical decentralization of information flow guarantees higher levels of freedom of expression. In fact, the Internet has been what activists turned to when they were denied access to other communication channels. According to the 2005 Human Rights Watch report, "Articles that had been censored in print publications such as the *Cairo Times* and the *Middle East Times* ran in their entirety online. Egyptian organizations and individuals continue to criticize government policies and individual officials in strong terms online" (p. 34). Indeed, Egyptian human rights activists have hailed the Internet as a medium that added a lot to the margin of freedom in the country, particularly with regard to political, religious, or human rights issues (HRW, 2005).

The Internet has enjoyed a good deal of freedom in Morocco. According to Human Rights Watch, there were no reports of government censorship of Web content or monitoring of e-mail or Web-based communications. Anyone can obtain an Internet account or put up a Web site without any government permission needed. Internet users can remain anonymous if they wish to. However, private ISPs have to provide access to users through Maroc Telecom (*Itisalat Al Magrib*), which competes with them as an ISP in its own right. In

November 2004, Morocco hosted a conference on freedom of expression as a follow-up to the first phase of the WSIS, held in Geneva in 2003. The conference spoke strongly against heavy regulation of the Internet, and stressed that national security and the fight against terrorism should not formulate an excuse to inhibit freedom of expression on the Internet or on any other media (RSF, 2004).

Bahrain is trying to use the Internet to enhance its economic situation. As a result, the government has tried to maintain a good degree of freedom to Internet users. Internet access has been available to the public in Bahrain since 1995. The government does not have any direct monitoring on online content, and does not require licensing for Internet access or Web site development. For citizens who would like to block some sites from their homes or their children, the government has offered free filtering software to be set up and used by the users themselves. The country has a sole ISP though, Bahrain Telecom (Batelco), a state-owned and operated company. The sole ISP makes it easy to monitor e-mail communication as well as to monitor content and detect any "inappropriate" material, although the government asserts it does not do that. There have been some reports though that some opposition political sites have been blocked at times, but those seem to be rare cases. Overall, the flow of information on the Internet in Bahrain seems to enjoy a decent degree of freedom (RSF, 2004).

Similarly, in Kuwait, the government does not directly interfere in information flow on the Internet. However, most ISPs go through the Kuwaiti Ministry of Communication, and so it is easier to have a proxy server with filtering software to monitor content. But so far, no reports are known of cases of Internet filtering or censorship. There are also other ISPs that use satellite connections to international Internet gateways outside Kuwait (RSF, 2004).

Finally, it is interesting to note that several Arab governments have started (or increased) their efforts to monitor information flow on the Internet after the attacks of September 11, 2001. Most of these efforts came in response to pressure from the United States government to crack down harder on terrorism (or terrorist communication) within the Arab nations, since it was reported after the September 11 attacks that the terrorists used e-mail as a main channel of communication (Franda, 2002). It is ironic that Arab governments impose stricter censorship and monitoring measures as a result of calls from the world's number one democracy advocate, the United States of America. Most Arab countries who use sophisticated filtering systems (e.g., Saudi Arabia and the United Arab Emirates) do so with the help of U.S.-made and/or U.S.-based software and proxy servers. Once these systems are in place, they are used to monitor not only alleged "terrorist" communications, but any other sort of content that such governments deem "inappropriate" for their citizens.

It is only appropriate to end this chapter by reiterating the recommendations of the Morocco conference on online freedom of expression. The fight against terrorism should not be taken as an excuse to restrict freedom of expression and opinion whether online or offline.

· 8 ·

IS THE INTERNET GOOD FOR YOU? PATTERNS OF INTERNET USE WORLDWIDE

The main study featured in this book was conducted by the author during March and April 2003 as the basis for her doctoral dissertation research at the University of Miami, Florida. The study is grounded in theory, and in particular, utilizes the theoretical approach of media uses and gratifications. Before going into the details of the study, I will present you with a brief account of the theory as well as a brief review of the relevant research on the Internet conducted in various parts of the world. In doing so, I will explain why I used this particular theory to conduct my research given the particular characteristics that the Internet possesses as a medium. In the following chapters, I will explain the scientific methodology and variables used in the study, before I go on to review the results.

Uses and Gratifications

The uses and gratifications approach is considered a theory of media effects, although it grew out of a frustration with the inability of media effects theories to arrive at conclusive evidence accounting for the effects of various media. The approach therefore decided to concentrate on the receiver of the communication message rather than on the message itself. Instead of studying what the message entails, the approach considers the receiver an active agent in the process of communication and argues that the audience is actively utilizing the media rather than being used or influenced by them. It therefore does not assume a direct relation between the communication message and the effects on the consumer. Rather, it argues that the consumer is the active agent that puts the messages to use, and however he/she uses the messages is an intervening factor that influences the effects of those messages on the individual. To give you an oversimplified example, the uses and gratifications approach assumes that if you watch three hours a day of television news, this will have a different effect on you than if you were to watch three hours of video clips every day. So it is you, the active audience, who chooses how to use television through your choice of content and, to an extent, what type of effect television may have on you. The uses and gratifications approach therefore argues that media use is an intervening variable that influences potential media effects on audiences.

Historically, research has always tried to utilize the uses and gratifications approach to gain a better understanding of how any new medium is used by its audiences. More recently, the approach has been applied to new media such as satellite television, video cassette recorders, remote control devices, mobile phones, and the Internet. The approach centers around the gratifications sought by the individual user. It takes into account antecedents to these gratifications, including such variables as social and psychological factors, media structure and availability, individual values and beliefs, and more recently, media technology.

The concern with social and psychological factors was one of the major steps taken by uses and gratifications researchers in way of the theoretical development of the approach. Researchers realized early on that the individual media user does not survive in a vacuum, but is rather affected by his/her social and psychological environment. E. Katz, Blumler, and Gurevitch (1974) suggested that "it is the combined product of psychological dispositions, sociological factors, and environmental conditions that determines the specific uses of the media by members of the audience." For example, people with "limited opportunities for social contacts" might be more likely to use media for companionship needs (p. 27). From a psychological standpoint, researchers presented general paradigms of human motivation, each of which could be used as a starting point for understanding why audiences seek a particular media gratification. The psychological paradigms were broken down by whether they were active or passive, internal or external, and cognitive or affective. They included needs of consistency, attribution, stimulation, identification, assertion, and tension-reduction.

So the uses and gratifications approach has as its starting point the consumer and the gratifications he/she seeks. It then moves on to the media or messages the consumer chooses, then the usage of those media or messages. It is only after considering usage that research can proceed to an understanding of effects on the audience. In this light, the argument holds that an understanding of the uses of a medium or a message is a prerequisite to an understanding of its effects on the audience. The best single sentence to describe the approach is probably still what E. Katz and Foulkes wrote in 1962, "This is the approach that asks the question, not 'What do the media do to people?' but, rather, 'What do people do with the media?'" (p. 378).

Unlike most media effects theories, the uses and gratifications approach assumes, first of all, that the audiences are active and goal-oriented, and that they are actually the ones to decide how to use the media in their lives. Second, it assumes that the audiences know what their needs are, and accordingly, they make a conscious choice of which medium or message to attend to at a particular time, and how to make use of it. Third, the media compete with other sources of need gratification since the audiences choose the medium to use from amongst an array of media and activities that could fulfill their needs. Fourth, from a methodological point of view, the approach assumes that many of the needs that audiences seek to gratify through mass media can be identified through data supplied by the audiences themselves. This assumes that the audiences are aware of why they use a certain medium or message, and that they can verbalize those reasons when asked, or at least they could recognize their reasons for use if confronted with them (E. Katz et al., 1974).

Alan Rubin, a leading uses and gratifications scholar, argued that since gratifications sought from a medium or message act as intervening variables to media effects, uses and gratifications research helped provide a map for different typologies of audience motives and media functions that conceptualized such gratifications. For example, A. Rubin reported that people who used television "to learn something" had a greater sense of the perceived reality of television content (A. Rubin, 1979). Palmgreen (1984) argued that audience gratifications were related to a wide variety of media effects, "including knowledge, dependency, attitudes, perceptions of social reality, agenda setting, discussion, and various political effects variables" (p. 29).

One of the issues that raised controversy in uses and gratifications research is the assumption that audiences are actually active. This assumption is put to a bigger test in societies that are still fighting illiteracy, such as the Arab world. Some research in the West has shown that audience members are active and do differentiate among media objects based on their gratifications sought and/or obtained. A. Rubin also looked at audience activity, and concluded that the concept should be viewed as a variable, since audiences were likely to fluctuate between passive and active at different times of their media use experience (A. Rubin 1983, 1984, 1985b, 1993, 1994). The Internet however seems to be a perfect medium for study under the active audience assumption, since it is not a medium that people display in the background like they do with television or radio. Rather, when people log on to the Internet, they are usually seeking a particular activity or a particular aim in mind (even if it is to kill time, they still know that this is why they are logging on).

Uses and gratifications research has always been popular whenever a medium is considered "new." In the last decade or so, researchers have used the approach to conduct research on newer media such as video cassette recorders (VCRs), cellular phones, satellite channels, remote control devices (RCDs), and finally, the Internet. Researchers have often argued that the approach is useful in investigating the relationship among people and different communication technologies. For example, Carolyn Lin (1992, 1994) examined the uses of VCRs. Her survey-based research looked at how the motives of VCR use, satisfaction, and interpersonal communication related to home entertainment, social utility, and displacement. The researcher further examined viewing attitudes and the emergence of a "home entertainment culture." Results showed that home viewing attitudes predicted the types and motives of VCR use.

Lin (1993a, 1993b) also looked at television viewing motivations in the new media context of multichannel viewing opportunities provided by cable and RCDs. She concluded that the multitude of options and viewing opportunities created by new media and RCDs "has created new audience activity dimensions" such as grazing, or remote control zapping to assess viewing options. The researcher also found that cable and RCDs gave viewers an enhanced sense of control over the environment, which related to stronger entertainment gratifications.

Special Characteristics of the Internet As a Medium

Being a relatively new medium, some researchers argued that the Internet stands to benefit from the uses and gratifications approach. Rayburn (1996) argued for a uses and gratifications perspective to studying the Internet. He noted that Internet consumption is especially purposive because of its interactivity. In his own words:

> While it would be naïve to assume that *all* media consumption is purposive, certainly it is safe to assume that some consumption is purposive, especially consumption of interactive media. Interactive media are, by definition, intentionally consumed as an audience member must make conscious choices about which path to choose (p. 157).

Ruggiero (2000) enthusiastically pushed for an exploration of the Internet from a uses and gratifications perspective. He asserted that the introduction and increasing popularity of computer-mediated communication (CMC) has revived the uses and gratifications approach. "In fact, uses and gratifications has always provided a cutting-edge theoretical approach in the initial stages of each new mass communications [sic] medium: newspapers, radio and television, and now the Internet" (p. 3).

Back in 1985, Williams, Phillips, and Lum argued that the uses and gratifications approach is especially useful in examining "new media." At that time, the new media they were referring to were "visible media as video cassette or disk, cable television, new telephone services, home computers, videotext or teletext services, and . . . the less visible technologies of digital switching, satellite transmission, and broadband telecommunications networks." The authors identified some of the technologically based characteristics of those new media as follows:

> Making distance all but irrelevant (communications satellite).
>
> Freeing television from the restrictions of broadcast schedules (videotape).
>
> Providing for nonlinear access to information (computers, video disks).
>
> Offering nearly unlimited availability of two-way voice or text communications (mobile telephone, computer teleconferencing).
>
> Transporting many simultaneous message or program choices (coaxial cable, fiber optics).
>
> Bypassing the printing and transportation requirements for the transmission of textual information (video and teletext) (p. 242).

Interestingly, the authors point out that,

> There is not so much that is "new" in the aforementioned technologies. To the user, most are merely extensions of existing media. . . . Most important . . . is that the new

> media offer many increased alternatives for both access to and interaction with message stimuli (Williams et al., 1985, pp. 241–242).

A decade later, however, a "new" medium was available that managed to encompass all the attributes of those other "not-so-new" media. The Internet does provide nonlinear access to information at any time of day or night. The millions of Web pages have all bypassed the traditional requirements of printing and transportation for the transmission of information. Users can choose to "tune in" to the Internet as a radio device by simply logging on to any online radio station, or watch their favorite video clips from a choice of more Web sites than they can keep track of (Abdulla, 2003). The Internet also makes distances all but irrelevant, in the sense that users can e-mail or chat with each other all over the world at any given moment. As Jones (1997) put it, the Internet allows for "interaction at a perfect closeness yet an infinite remove, keeping us 'in' touch but unable *to* touch" (p. 26, emphasis in original). In a sense, it actually makes the concept of time a little fuzzy, as a user can chat with another synchronously across several time zones. Consider, for example, a user who logs on at midnight in California to chat with his/her friend in Cairo, where it is 10 a.m. It is literally daylight at one user's end, and nighttime on the other end of the line (Abdulla, 2003).

Perse (2001) noted that uses and gratifications research has shown that motives and use patterns influence potential media effects, and that motives for using new technologies should be investigated in relation to each technology's particular characteristics. Internet characteristics that have been identified by researchers include interactivity, demassification, multimedia, hypertextuality, synchroneity, and asynchroneity. Interactivity in communication is mostly attributed to interpersonal communication and defined as the extent to which participants of a communication act can control and exchange roles in a mutual discourse (Morris & Ogan, 1996; Ruggiero, 2000). More recently, it has been applied to forms of new media such as two-way cable or interactive television. In a more computer-mediated communication sense, interactivity is the degree to which the technology is capable of instantly responding to user demands.

Ruggiero (2000) argued that interactivity is especially important to applying uses and gratifications since the approach has at its heart the notion of an active audience. Dyson (1993) concluded that interactivity means more to Internet users than multimedia videos and online shopping. It might actually enable them to create new means of communication. Morris and Ogan (1996) reminded us that it is useful to think of interactivity as a variable rather than a constant, since it changes with the nature of the particular Internet task or function in use at any given time.

Another characteristic of the Internet is its ability to act both as a mass and a personal medium, what researchers have termed "demassification." Williams, Rice, and Rogers (1988) defined demassification as the extent to which an individual user has control over a medium, making it more and more similar to interpersonal communication. New media like the Internet provide a much bigger choice to users, from which they can select and "personalize" their choices to fit their needs and likings. Most Web-based news sources, for example, offer an option for users to create a personalized Web page that features only the news of their choice. Cable News Network (CNN) was among the first news networks that provided this service. Today, most news sites and most commercial sites and portals provide a similar service. For example, My Yahoo! can carry items of the user's individual choices from news to online shopping to health to travel. Arab sites such as Masrawy.com and Arabia.com now offer the same services. Sites such as Amazon.com and BMG Music provide personalized Web pages that will recognize the IP (Internet Protocol) address of a particular computer, so that the moment the user logs onto the site, the screen will flash a personalized greeting with the user's name. Such pages also carry the site's recommendations for best buys that fit the user's profile, psychographics, and personal taste (based on previous shopping patterns and on information the user provided when customizing the page). This demassification of the medium, like interactivity, is especially relevant to the application of the uses and gratifications approach, since a basic assumption of the approach is that audiences choose from a large variety of media. The Internet has offered more choice options than was ever available before (Abdulla, 2003).

The Internet is also considered to be both synchronous and asynchronous in nature, depending on the particular Internet application or service in use. For example, e-mail is considered asynchronous because users can send and/or receive messages at different times at their convenience. The sender and receiver do not have to be online at the same time. At the same time, other Internet applications such as chatting and instant messaging are synchronous in nature. Users interact with each other instantly across different borders and different time zones (Abdulla, 2003).

Internet Uses and Gratifications Research

Some studies, mostly in the United States, have examined the Internet from a uses and gratifications perspective. Most of these studies focused on students' use of the Internet. T. Perry, L. Perry, and Hosack-Curlin (1998) surveyed students on three college campuses to assess their patterns of Internet use. They reported that e-mail was the most popular Internet function among students, with about 80% of the students using it. E-mail was particularly popular with students 18–19 years of age, with a reported usage rate of over 90%. Students seemed to regard the Internet as an important medium, with 93% saying they would use it in their future careers, and 95% saying the Internet is not a "passing fad."

Perse and Ferguson (2000) conducted a Web-based survey of 250 students to assess the perceived benefits that audiences derive from Web use. The findings of the study indicated that learning was rated the most salient benefit of Web use, followed by pass time and entertainment. The researchers found support for the impact of perceived benefits and costs on Web surfing. Web sites about entertainment and sports were the ones most likely to be linked to beneficial Web use.

Lin (1999, 2001) looked at factors that predicted online service adoption. This study could be of particular interest to us in the Arab world since we do have a problem of low adoption figures. The results indicated that the strongest predictors of Internet adoption were cognitive and affective gratifications for the online service. The research also indicated that adopter attributes were moderate predictors, and that other media use attributes were weak predictors of potential online service adoption.

Perhaps the most relevant study to the one I conducted from a methodology point of view is the one conducted by Papacharissi and A. Rubin, published in 2000. The researchers surveyed 279 students to look at the way they use the Internet through the uses and gratifications approach. The researchers found five primary motives for using the Internet. Those were, in order of importance: information seeking, entertainment, convenience, passing time, and interpersonal utility. The study also examined behavioral and attitudinal outcomes of Internet use such as amount of Internet use, duration of Internet use, Internet affinity, and Internet satisfaction.

Interestingly enough, the study concluded a contrasting pattern between the concepts of Internet affinity and satisfaction. Those who used the Internet for interpersonal utility did not necessarily feel more satisfied about this choice, and those who used the Internet to obtain information did not actually consider it to be really important in their lives. Satisfaction was associated with a more instrumental approach to the medium, whereas affinity was linked to a more ritualized use of the Internet (Papacharissi & A. Rubin, 2000).

A few studies attempted to look at Internet use in Asia. Chao (2001) looked at instructional uses of the Internet and perceptions of their benefits and limitations among university students in Taiwan. She surveyed a cluster sample of 527 undergraduate students enrolled in general education classes at two universities. She reported that the majority of Internet users in her sample (97.6%, n = 487) used Chinese as their main language on the Internet. The majority of the sites her respondents reported accessing on the Web were "within Taiwan" (97.4%, n = 486) as opposed to in "other countries." Students' perceived barriers to Internet use included "Internet traffic congestion" (79.7%, n = 372), not finding desired information (73%, n = 341), as well as the "lack of English skills in reading and writing on the Internet" (69.8%, n = 326). "Lack of adequate knowledge or skills to use the Internet" was reported by 55.9% (n = 261), while cost was reported by 36.8% (n = 172).

Choi (2001) investigated Internet use patterns and motivations among Korean users. Combining uses and gratifications with expectancy-value theory, Choi administered an online survey through Korean portal sites. The findings of the study, based on 355 responses, indicated that Koreans use the Internet for social interaction, entertainment, sexuality, and online transactions. Other motivations such as information seeking, escape, and loneliness were not significantly related to Koreans' Internet use. The study showed that speed of Internet access was a significant factor in determining Internet use. For example, over 50% of respondents had fast access to the Internet, and 70% of those fast access users accessed online games, as opposed to 38% of the regular dial-up access users.

Choi (2001) also found that heavy Internet users (more than 140 minutes per day) were more likely to access the Internet for entertainment, online transactions, and social interactions. They were also more likely to use the Internet to access discussion boards and chat rooms, to play games and to seek sexual materials. Korean males were more likely than females to be heavy Internet users. They were also more likely to seek sexual motivations on the Internet, as well as to play online games. Females were more likely to use the Internet for social interaction

Does the Internet Promote Loneliness or Depression?

One of the controversial points that were raised concerning the Internet's effect on youth is the potential of the Internet to draw people into its world and out of the world of social, educational, or family activities, and the possibility of this promoting loneliness or even depression among these users. The study that gave rise to the controversy was conducted by Kraut, Lundmark, et al. in 1998. They conducted a longitudinal study, whereby they examined Internet use patterns among 73 families (169 individuals) over their first one or two years of Internet use. Measuring dependent variables before and after the Internet came into their respondents' lives, the researchers found a weak but significant relationship between Internet use and loneliness ($r = .15$), and between Internet use and depression ($r = .19$). They also reported that increased Internet use was associated with less family communication, social support, and face-to-face social interactions.

In 2000, Nie and Erbring published another study that supported these findings. Although the researchers emphasized that their results are preliminary, they argued that, "the more hours people use the Internet, the less time they spend with real human beings." The study also found that Internet users spent less time watching television or reading newspapers, although the study acknowledged that this might be because people are reading the news online. The researchers concluded that, "The more time people spend using the Internet, the more they turn their back on the traditional media."

Several points need to be made regarding these two studies. First of all, they are only two studies among a multitude of studies, most of which arrived at very different results than these two (some of which are reviewed below). Second, the correlations that these studies established, although statistically significant, were very weak by any scientific measure, that they almost do not mean anything. Third, there are many intervening variables that could have confounded the results of these two studies. The researchers did not look at the psychological state of their respondents independent of their Internet use and did not investigate whether any other life events or factors could have caused the change. It is therefore very possible that these very weak correlations were established because the respondents who reported them happened to have a bad day when they were interviewed, or had been through a personal trauma that caused their loneliness or depression that had absolutely nothing to do with their patterns of Internet use. It is scientifically impossible to assume a cause-and-effect relationship between Internet use and loneliness or depression based on these studies.

As I mentioned before, other studies were conducted to further examine the social and psychological aspects of Internet use. Flaherty, Pearce, and R. Rubin (1998) looked at CMC as a functional alternative to face-to-face communication. In other words, they wanted to know whether people who use the Internet feel it is a valid alternative to interacting with their families and friends. The researchers surveyed a sample of 132 Internet users to examine the relationship between motives for using the Internet and motives for face-to-face communications. Results showed that communication motives for the two modes of interaction are in fact different. Internet use as a communication channel was not perceived by respondents to be a functional alternative to face-to-face communication with family and friends.

Another study was conducted by J. Katz and Aspden in 1997. The researchers investigated patterns of participation in community, leisure, and political activities among Internet users. They conducted a phone survey of a random sample of 2,500 adults, plus another random sample of 400 Internet users. They found no significant differences between Internet users and non-users regarding such activities, or regarding time spent with family and friends. Furthermore, respondents reported that the Internet helped them keep in touch with distant relatives and meet new friends. Other studies by the same researchers showed a positive correlation between Internet use and patterns of interpersonal communication, that is, the more people used the Internet, the more they also interacted with family and friends. Internet use was also associated with greater levels of telephone use and social interaction. Respondents also reported developing face-to-face friendships out of their online interactions. The researchers concluded that, "Far from creating a nation of strangers, the Internet is creating a nation richer in friendships and social relationships," (J. Katz & Aspden, 1997a, 1997b, 1997c).

Other research focused on the Internet as an alternative to other media. One such study conducted through a U.S. national survey reported that contrary to expectations, Internet use is not likely to be the cause of decline in the use of other media. Internet users were more likely than nonusers to read newspapers and listen to radio news. As for television news viewing, there was no significant difference between users and nonusers of the Internet (Stempel, Hargrove & Bernt, 2000).

Similarly, Weaver (1998) found that Internet use does not replace use of other media, with the possible exception of television. The study found that users of the Internet are more likely to read newspapers and print media than nonusers. Internet users were also more likely to go to the movies and listen to the radio than nonusers.

Weiser (2000) reviewed the studies linking Internet use to negative social and psychological factors. The researcher concluded that "the intuitive assertion that increased Internet use has uniformly negative social, psychological, and interpersonal consequences is premature and perhaps a bit misleading. . . .

Internet use, if anything, increases psychological well-being, and facilitates the development of on-line relationships, which become more like real world relationships through time" (pp. 142, 143). The researcher also conducted an online survey to assess the psychological consequences of Internet use. He found that, contrary to Kraut, Lundmark, et al. (1998), increased Internet use has "very weak direct relations to psychological well-being" (p. 257). The researcher's conclusion again points to the importance of studying the uses and gratifications of the Internet. He says, "It appears as though the psychological effects of Internet use depend primarily upon the functions which it serves for users, and how these functions ultimately affect one's level of social integration" (p. 257). He concluded however that practical use of the Internet may have a beneficial effect on psychological well-being.

Internet Research in Egypt and the Arab World

Wheeler (1998) wrote a qualitative article on Internet use and practices in Kuwait. Through ethnography and participant observation, the researcher posed some insights about what she called "the development of an Internet culture in Kuwait." The researcher noted that while Kuwait is promoting Internet use, it is worrying about "the effects of foreign values on society." Through investigating the practices of the Kuwaiti government, as well as companies and individual citizens, she noted that the notion of cyberspace and its uses was bound to differ based on the local culture. Throughout her article, she argued that information is regarded as a threat rather than an opportunity in Kuwaiti society, and that there is a feeling among citizens that having and/or verbalizing an opinion, particularly a political opinion, is bad. She concluded that local identity and cultural frameworks played a vital role in the way and the degree to which a society would adapt to a global economy or a global culture.

Kuwaiti academicians (Abbas, 2001; Abbas, 1997; Al-Mazeedi & Ismail, 1998) reported that students constitute the majority of Internet users in Kuwait, as well as the group most likely to be affected by the Internet. The researchers all agreed that the Internet is presenting a new window onto the world to Kuwaiti students, although it might not be in strict alignment with the traditional tenets of their culture (cited in Wheeler, 2003). Females interviewed by Wheeler (2003) said that they used the Internet to communicate with the opposite sex, and found it to be educational because "they did not generally have first-hand knowledge of how men think." Those females added that they valued the experience because it taught them about the opposite sex within the safety of being confined to a computer screen. Wheeler concluded that "the most magnetic quality of the Internet, which draws Kuwaiti youths to the technology, is the way in which it enables the transgression of gender lines, which are otherwise strictly enforced in Kuwaiti society," particularly among youths who went to segregated schools for most of their lives. She argued that such communication could lead to a better understanding of the other sex in general, and therefore to better marriage and family relationships in Kuwait. She also noted that the traditional and Islamic cultural norms that Kuwaiti students grow up with ultimately shape their long-term Internet use, so that they refrain from usage patterns that strongly contradict these values, such as accessing pornography sites. Instead, even if they experiment, they eventually adapt their patterns of Internet use to their upbringing, and the societal norms and values they cherish.

A similar research article (Piecowye, 2003) was conducted on female students in the United Arab Emirates. The findings contradicted the notion that the Internet will eventually reshape Arab cultures by Western influences. Through brief interviews with 22 students at Zayed University, the researcher noted that his students were active audiences of the Internet, who knew exactly what aspects of the global cultures they wanted to expose themselves to and what aspects they turned away from. Through e-mail exchanges with students at a London-based university, students used the opportunity to share their Arab culture and tradition, while insisting on preserving this culture. The researcher concluded that the Internet is more likely to reinforce than challenge the Arab culture, while providing an opportunity for sharing it with the world. Intercultural connections, he argued, force Arab students to think about their culture and their values, and serve to define those values in the context of other global cultures. He noted that technology was serving to extend the traditional practices of UAE culture, as the e-mail exchanges were almost a virtual extension of the *majlis*, the traditional meeting environment where people gather to talk and exchange ideas. He concluded that while the United Arab Emirates cannot ignore the Internet as a technological revolution, female Emirati students are not the helpless victims of globalization. Rather, they are conscious media consumers who are capable of determining what aspects of the local and global cultures they can accept, preserve, share, or reject.

Wheeler (2004) also examined the female culture of the Arab world and some of its patterns of Internet use. She argued that women stand to gain much from the Internet in terms of general empowerment and as an educational and professional tool. The researcher conducted interviews with women in Egypt,

Morocco, Saudi Arabia, Kuwait, and Bahrain to find out how and why they use the Internet, and what this new medium means to them as Arab women. She reports some varying patterns among users. For example, while the females she interviewed in Egypt mostly used the Internet for work-related aspects, this was not the case in Kuwait where females used it mainly as a means of socializing and entertainment. The researcher gave a cultural explanation for this difference, pointing out the fact that in Kuwait, there is segregation of sexes, and therefore the Internet provides a virtual space for meeting members of the opposite sex.

In interviews with women from Morocco and Saudi Arabia, the issue of censorship comes up. In Morocco, the Internet seems to provide a way for some groups (the researcher mentions in particular *Al-Adl Wa'l Ihsan*[1] opposition group) to circumvent the government censorship on traditional media by going online. In Saudi Arabia, as we mentioned in chapter 7, Internet users sometimes access the Internet by connecting to an ISP in Bahrain to avoid Saudi government censorship on online materials. This was especially true before 1999, when there was no Internet in Saudi. But even now, Saudi citizens are willing to pay for an international phone call to Bahrain through which they can access the Web without government restriction (Wheeler, 2004).

In Egypt, a few studies have examined some aspects of the Internet culture. As early as in 1996, one study (W. Hassan, 1996) looked at the prospects and concerns of the Egyptian information superhighway. The researcher surveyed a purposive sample of media professionals and pointed out to the importance given by those professionals to the potential effects of the Internet and other new information technologies on the Egyptian society and the Arab world. Such effects included the introduction of online journalism, digital imaging, and digital cable. He argued that this would lead to accelerating the exchange of information and modernizing the media systems in the Arab world. Social effects would include more exposure to world cultures, and more democratization of Arab political systems. Concerns identified at the time included affordability and financial aspects, training of personnel, and the availability of the technological facilities needed.

Ibrahim (1999) looked at Internet use among journalists in Egypt. He pointed out the problem of as yet limited Internet access. At that time, journalists using the Internet were mainly confined to those concerned with foreign affairs, translation, and technology. He concluded that the Internet at that time was used as a means of gathering information more than anything else. Abdel Salam (1998) pointed out that Egyptian and Arab newspapers and magazines simply publish the same content online, instead of taking advantage of the interactive nature of the Internet, a problem that is still just as valid today as it was in 1998.

Another study (El Hoteiby, 1999) looked at online publishing in Egypt and the Arab world. He conducted an online survey to a purposive sample of editors of Arab online newspapers and magazines (that also have print editions). The results indicated that respondents believed online journalism would give them access to more freedom of expression and more flexibility in news gathering as a result of the multitude of information sources available on the Internet. Reporting techniques would be enhanced as a result, and the standard of Arab journalism would be elevated. This would eventually enhance the information exchange, and create a more democratic civic society.

Nagib (2000) examined the impact of Arabization. She anticipated that as the use of Arabic characters on the Internet increases, the number of Arab users is likely to multiply since they will no longer have to deal with the language barrier. Today, Arab users are utilizing new features on the Web such as Arabic-language e-mail access provided by Arabic portals such as Arabia.com, Maktoob.com, and Masrawy.com. As Nagib put it,

> The realization that Arabs have a unique identity and a desire to reach out to each other in their own mother tongue has impacted the development of the Internet. Nations are realizing more and more each day that they have a need to express themselves in their own language to help them express themselves freely using their own dialect (p. 144).

As we will discuss in the findings and the recommendations of the main study in this book, the use of language, Arabic and Arabic in Latin characters, is an all important and interesting topic. Arabic content on the Web should increase at a much higher rate than it is now. Egypt should be among the leading countries to develop such content, given its history in media leadership and the abundance of computer and software programing experts in the country.

The Internet provides a rich opportunity for communication researchers to step in and assume an active role in examining its different aspects as a new medium of communication. Certain characteristics of the Internet, such as interactivity and demassification, are right at the heart of the basic assumptions of the uses and gratifications approach, namely, the concepts of audience activity and choice. The Internet offers choices for audiences that exceed those offered by any other medium. In fact, the Internet seems to offer a bouquet of media in one. Communication researchers should actively engage in more research on this rich medium to help us understand the different motivations behind its use. And this is what I will try to explore in the remainder of this book.

· 9 ·

THE USES AND GRATIFICATIONS OF THE INTERNET AMONG ARAB STUDENTS IN EGYPT: METHODOLOGY

The study I conducted investigated the uses and gratifications of the Internet among students in Egypt. It surveyed Arab English-speaking university students at the American University in Cairo. The main questions addressed were how university students used the Internet and what gratifications they derived from this new medium of communication. Variables investigated in relation to the uses and gratifications approach were gender, year in school, type of Internet connectivity, level of skills needed, Internet exposure, Internet affinity, and Internet satisfaction in an effort to determine what predicts Internet use and Internet exposure among Arab students in Egypt. The study also explored patterns of other media and technology ownership among Internet users, and discussed the potential effects, if any, of Internet use on patterns of social relationships and media consumption.

As I mentioned in the previous chapters, the study of the uses and gratifications of the Internet is active in the United States, Asia, and Europe. However, such basic knowledge of how people use the Internet is rare in the Arab world. This study is one of the first to assess such information. The knowledge of how people use a particular medium is an important step on the road to more knowledge about the developments and the potential effects of the medium. With the Internet becoming so popular, and with all the potential that this new technology provides, knowledge of how people use it is especially important. The Internet could be of particular importance to the Arab world since it provides a tool for free expression and mobilization of civil society in such a way that has never been available before to this part of the world. An investigation of how the Internet is used by Arab students provides valuable information about what goes on in this part of the world. This information is crucial to providing more understanding of the Arab culture, how technology impacts it, and how it impacts technology consumption patterns. It is also important in assessing the potential effects of the Internet on this part of the world.

Research Questions and Hypotheses

Based on the literature review, the following research questions and hypotheses were formulated for this study.

The main concern for uses and gratifications research has been to investigate the uses of a particular medium or message by the audiences (see, for example, Charney & Greenberg, 2002; Clifford, 1995; Ebersole, 2000; Greenberg, 1974; Jeffres & Atkin, 1996; Leung & Wei, 2000; Lin, 1992, 1994, 1999, 2001; Maignan & Lukas, 1997; Palmgreen & Rayburn, 1979; Papacharissi & A. Rubin, 2000; Perse & Dunn, 1998; Perse & Ferguson, 2000; A. Rubin, 1977, 1983, 1985b; Schramm, Lyle, & Parker, 1961; Wells, 1990). The following research question was therefore advanced:

RQ1: *How do Arab English-speaking university students in Egypt use the Internet?*

Some studies found an association between media use patterns and demographic factors, particularly age and gender (see, for example, Abdulla 1996, 2000; Elton & Carey, 1983; Greenberg, 1974; Mesbah, 1991; A. Rubin, 1977, 1979; A. Rubin & R. Rubin, 1982; Schramm et al., 1961; Wells, 1990). Others have specifically investigated the relationship between demographics and Internet use (Choi, 2001; Kraut, Mukhopadhyay, Szczypula, Kiesler, & Scherlis, 1998). Since this study is concerned with students who are likely to be in a limited age bracket (mostly 18–22), only gender will be investigated.

H1: *There is a significant difference in Internet use motives between males and females among Arab English-speaking university students in Egypt.*

H2: *There is a significant difference in Internet exposure in terms of time spent on the Internet between males and females among Arab English-speaking university students in Egypt.*

Mesbah (1991) reported a positive relationship between education level and more intentional television viewing. Atkin, Jeffres, & Neuendorf (1998) found a positive relationship between education and Internet adoption. It is logical to assume that as students spend more time in a college environment, they are more likely to be introduced to and be more familiar with the Internet. The following research question was therefore advanced:

RQ2: *Is school year a significant factor in how Arab English-speaking university students use the Internet?*

Some studies looked at the relationship between Internet use and affinity, satisfaction, skill level, and amount of Internet exposure. Ebersole (2000) found that Internet affinity and skill level correlated positively with motivations for Internet use. Papacharissi and A. Rubin (2000) reported that Internet affinity related more to ritualized Internet use, and that Internet satisfaction related to instrumental uses. They investigated whether Internet use motives predicted time spent on the Internet, but found no statistical significance. They did not look, however, at how operational skill levels affected Internet exposure. Chao (2001) found a positive correlation between Internet operational skills and Internet use for educational purposes in Taiwan. The following research hypotheses were therefore advanced:

H3: *There is a significant positive correlation between Internet affinity and Internet use motives among Arab English-speaking university students.*

H4: *There is a significant positive correlation between Internet satisfaction and Internet use motives among Arab English-speaking university students.*

H5: *There is a significant difference between high and low levels of operational skills necessary for Internet use (typing speed, English-language proficiency, Internet use proficiency) and Internet use motives among Arab English-speaking university students.*

H6: *There is a significant difference between high and low levels of operational skills necessary for Internet use (typing speed, English-language proficiency, Internet use proficiency) in terms of time spent on the Internet among Arab English-speaking university students.*

Swanson (1987) argued that attempts to further strengthen the theoretical grounding of uses and gratifications should focus in part on the relationship between gratifications and media exposure. Choi (2001) found that Koreans who have fast access to the Internet used it differently than those who had regular dial-up connections. He also found significant differences between heavy and light Internet users in terms of their Internet use motivations. The following research hypothesis and questions were therefore set forward:

H7: *There is a positive correlation between Internet exposure and Internet use motives among Arab English-speaking university students.*

RQ3: *Is type of Internet connection (dial-up modem versus fast access) a significant factor in how Arab English-speaking university students use the Internet?*

RQ4: *Is type of Internet connection (dial-up modem versus fast access) a significant factor in the amount of time Arab English-speaking university students spend on the Internet?*

Some research has tried to establish whether or not Internet use affects patterns of use of other media and other activities and social interactions. The findings have not been conclusive. While some studies (Coffey & Stipp, 1997; Kraut, Lundmark et al., 1998; Nie & Erbring, 2000; Weiser, 2000) argued that Internet users spend less time on traditional media and other social activities, others (Atkin et al., 1998; Ferguson & Perse, 2000; Stempel, Hargrove, & Bernt, 2000; Weaver, 1998) reported that Internet use is not likely to cause decline in the use of other media, with the possible exception of television. Research also showed that the Internet is not likely to be an alternative for face-to-face communication or social interactions with family and friends (Flaherty, Pearce, & R. Rubin, 1998; J. Katz & Aspden, 1997b). Moreover, some studies (J. Katz & Aspden, 1997a, 1997c) reported increased levels of social interactions with more Internet use. The following research questions were therefore posed:

RQ5: *How does Internet use affect use of other media among Arab university students?*

RQ6: *How does Internet use affect social activities among Arab university students?*

One of the more sophisticated goals of uses and gratifications research has been to predict patterns of media use among audiences (see, for example, Choi, 2001; Conway & A. Rubin, 1991; Ebersole, 2000; Jeffres & Atkin, 1996; Leung & Wei, 2000; Lin, 1992, 1994, 1999, 2001; Palmgreen & Rayburn, 1982; Papacharissi & A. Rubin, 2000; Perse & Dunn, 1998; Perse & Ferguson, 2000; A. Rubin, 1985b; Wells, 1990). This study helps us explain the variables that predict Internet use and Internet exposure (time spent on the Internet) among our target population.

RQ7: *What variables predict Internet exposure among Arab English-speaking university students in Egypt?*

RQ8: *What variables predict Internet use among Arab English-speaking university students in Egypt?*

Sampling

Drawing a probability sample of Internet users is one of the biggest problems facing Internet researchers. Ruggiero (2002) pointed out the "practical impossibility of probability sampling on the Internet" (p. 13). Kaye and Johnson (2002) argued that since obtaining a probability sample is extremely problematic on the Internet, a purposive sample is sometimes better suited to the nature of the study. Researchers have often used a purposive student sample for Internet research, and several have argued for the benefits of using such a sample, since student populations usually have free access to the Internet through their schools and are likely to use it (see, for example, Charney & Greenberg, 2002; Chao, 2001; Ebersole, 2000; Ferguson & Perse, 2000; Kandell, 1998; Papacharissi & A. Rubin, 2000; T. Perry, L. Perry, & Hosack-Curlin, 1998; Perse & Ferguson, 2000; Scherer, 1996). This study used a Web-based, anonymous, online survey, the link to which was e-mailed to the population of undergraduate and graduate student Internet users at the American University in Cairo (AUC), Egypt.

Although AUC students are not necessarily typical of Egyptian or Arab universities' students, they do represent the population of student Internet users at other Egyptian and Arab universities. The more than 5,000 AUC students all speak English (English is the official language of instruction at AUC), and students are generally classified as upper-middle to upper socioeconomic class, making them capable of affording a computer and Internet connectivity.

This population was also chosen for practical reasons, since the percentage of students using the Internet at AUC (estimated at about 90%) is much larger than that at Cairo University (estimated at about 3–5%) (ACS, 2002). Cairo University is the largest Egyptian university, with a student population of over 200,000. It is also the most Internet-connected among public universities. However, with a 3–5% user estimate, it would have been practically impossible to locate a sample of Internet-user students to complete the survey. In addition, there is no plausible way of obtaining e-mail addresses for such a student population, which makes their participation in an online survey almost impossible. With the large growth rate of the Internet in Egypt (500% in 2002–2004), AUC students represent the typical demographics for Egyptian and Arab university students who are likely to be using the Internet at this time, or to adopt using it within the next few years. From a practical point of view, AUC students also offered a manageable population for sampling purposes (Table 9.1).

Table 9.1. The American University in Cairo Profile (2002–2003)

Number of enrolled students	
Undergraduates	4,191
Graduates	811
Nondegree	258
Total	5,260
Gender	
Male	47.5%
Female	52.5%
Nationality	
Egyptian	86.2%
Non-Egyptian	13.8%
Faculty	
Full-time	307
Part-time	228
Total	535
Faculty to student ratio	1:12.7
Recent graduation patterns	
2000	912
2001	796
2002	884

Note. From AUC Office of Planning and Institutional Research (2002).

AUC offers a free e-mail account on its server (username@aucegypt.edu) to all members of its community—students, faculty, and staff. Students can keep using their AUC e-mail accounts free of charge even after they graduate. However, not all students choose to use an AUC e-mail account. Some students opt to use or continue using other e-mail accounts, either Web-based such as Hotmail or Yahoo! or through one of Egypt's many ISPs. AUC does not require students to have an AUC e-mail account, or to register a different account on its mailing list. As a result, the university's central e-mail system can only reach those who chose to sign up for an AUC account. There were no exact figures available at the time the study was conducted as to the number of enrolled AUC students who did have an active AUC e-mail account. However, they were estimated by AUC's Office of Academic Computing Services to be about 30–40% of the AUC student population, that is, about 1,578–2,104 students.

An e-mail invitation to the Web-based survey was sent out through AUC's central e-mail server to all enrolled undergraduate and graduate students who had an AUC e-mail account. The e-mail introduced the study and the researcher, stressed on the importance of the study, and asked students to click on the study link to fill out the questionnaire. It also encouraged them to forward the link to their AUC colleagues who may not use an AUC e-mail account. A second e-mail was sent to all students two weeks later. It thanked those who had already filled out the survey and reminded those who have not to please do so. The e-mail provided a link to the study and again asked students to forward the link to their AUCian friends. A third and final e-mail was sent a week later to the same effect. All in all, the survey was available to students for 28 days.

In order to increase the chances for students who have non-AUC e-mail accounts to participate in the survey, an announcement about the study ran for three weeks in *The Channel*, a widely read bi-weekly announcement bulletin that is distributed at all AUC gates and in all AUC offices. The announcement said, "Internet Research: Your participation is needed in an important study about Internet use patterns in Egypt and the Arab World. If you are an Egyptian or Arab student, please fill out the survey at (survey URL)." Students who accessed the link were taken to the cover letter. Another e-mail was sent to all AUC faculty members who had an AUC e-mail account explaining the study and asking them to encourage their students to fill out the survey or to forward the link to them if possible.

At the end of four weeks of posting the survey online, 502 responses were received. It is hard to calculate an exact response rate since it is not clear exactly how many students received the invitation to take the survey. However, based on the figures provided by AUC's Office of Academic Computing Services, the estimated response rate for the e-mail invitation sent to all students is between 23.9% and 31.8%.[1]

Operationalization and Measurement of Variables

For the purposes of this study, an Arab is defined as a citizen of a country that belongs to the Arab League of Nations *and* whose official language is Arabic. These countries are Algeria, Bahrain, Egypt, Iraq, Jordan, Kuwait, Lebanon, the Libyan Arab Jamahiriya, Morocco, Oman, the Occupied Palestinian Territories,

Qatar, Saudi Arabia, Sudan, Syrian Arab Republic, Tunisia, the United Arab Emirates, and Yemen. An Internet user is defined as a person who has accessed the Internet at least once during the past month. Since the study uses an online survey, the mere act of accessing the survey makes respondents eligible for participation as "Internet users." All variables in this study are self-reported.

The following variables are identified for this study:

Year in school: Defined as an ordinal variable (freshman, sophomore, junior, senior, graduate).

Gender: Defined as a nominal variable (male, female).

Internet exposure: Defined as a ratio variable in terms of hours and minutes spent on the Internet during an average week.

Type of Internet connection: Defined as an ordinal variable (dial-up modem to fast access).

Dial-up modem (of any speed).

Fast access broadband Internet connectivity (including DSL, cable, satellite, ISDN, network, or leased lines).

Internet operational skills: Three questions were devised to generate self-assessment of students' Internet operational skills, namely, level of Internet use skills, typing skills, and English-language proficiency. Skill level within each of these variables were defined as follows:

Internet use skills:
Beginner: I can barely do what I want.
Intermediate: I know my way around to the basics.
Expert: I know my way very well on the Internet.

Typing skills:
Slow: Under 25 words per minute.
Moderate: 25–50 words per minute.
Fast: Over 50 words per minute.

English-language proficiency:
Low: Okay. I can manage to write and have a conversation, but with some difficulty.
Moderate: Fair. I can write and converse on most topics, but I need to practice.
High: Fluent. I can write and converse on any topic, no problems.

Internet affinity: The degree of perceived importance of the Internet as measured by an Internet Affinity Scale. This 5-item Likert-type scale was modified from A. Rubin's (1981) Television Affinity Scale. It was used in Papacharissi and A. Rubin (2000) and has an established Cronbach alpha of .84.

Internet satisfaction: Following Palmgreen and Rayburn (1985) and Papacharissi and A. Rubin (2000), satisfaction with Internet use is measured by a Likert-type scale item asking respondents to indicate how much they agreed or disagreed with the following statement: "Overall, I am satisfied with the job the Internet does in providing me with the things I am looking for."

Motives for using the Internet: Defined as the reasons why a person uses the Internet. These are measured by a Likert-type scale at the interval level, based on Greenberg's (1974) and A. Rubin's (1977, 1979) Television Viewing Motives Scale, and Papacharissi and A. Rubin's (2000) Computer-Mediated Communication Motives Scale.

The Survey Instrument: Scales, Reliability, and Validity

The survey used an online Web-based questionnaire that consisted of closed-ended questions (except for asking for an estimate of average time spent on the Internet, how long the respondents have been using the Internet, and the year the respondents were born). Following the advice of Dillman (2000), the first part of the questionnaire consisted of general questions to ease in respondents. The questions were about where respondents accessed the Internet, how much attention they devoted to what they were doing, and how long they spent on different functions of the Internet (e-mail, Web browsing, chat, other) during an average week.

The main scale of the questionnaire measured motives for using the Internet. The scale was based on Greenberg's (1974) and A. Rubin's (1977, 1979, 1981) Television Viewing Motives Scale. It was modified to measure Internet use in Papacharissi and A. Rubin (2000) under the name Computer-Mediated Communication Motives Scale. The 5-point Likert-type scale asks respondents how much they identify with the reasons listed for using the Internet. The instructions used in this study stated, "Following are some reasons people give for using the Internet (including the Web, e-mail, chat/instant messaging/discussion boards, and other functions). Please indicate how much each reason is like your own reasons for using the Internet by filling in the appropriate bubble. "I use the Internet . . ." The scale lists 40 items and asks respondents to rate their answers on each item as "strongly agree," "agree," "neutral," "disagree," or "strongly disagree."

Reliability of motive dimensions has been established in many studies (Conway & A. Rubin, 1991; Finn & Gorr, 1988; Greenberg, 1974; A. Rubin, 1977; A. Rubin & Perse, 1987a) both in terms of inter-item correlations and Cronbach alphas (cited in R. Rubin, Palmgreen, & Sypher, 1994).

The motives scale was adapted to the Internet world by Papacharissi and A. Rubin (2000) under the name Computer-Mediated Communication Motives Scale. The researchers started out with 45 a priori items on the scale. After conducting a factor analysis, the researchers ended up with 27 items, loading "cleanly" on five factors.

1. Interpersonal utility: With an eigenvalue of 8.14 and a Cronbach alpha of .93, this factor included items that described categories of inclusion, affection, social interaction, expressive need, and surveillance. This factor accounted for 18.1% of the variance after rotation on a hierarchical regression analysis.
2. Pass time: With an eigenvalue of 3.38 and a Cronbach alpha of .85, this factor accounted for 7.5% of the variance.
3. Information seeking: With an eigenvalue of 3.73 and a Cronbach alpha of .87, this factor accounted for 8.3% of the variance.
4. Convenience: With an eigenvalue of 2.79 and a Cronbach alpha of .78, this factor represented categories of time control, convenience, economy, and social interaction, and accounted for 6.2% of the variance.
5. Entertainment: With an eigenvalue of 1.91 and a Cronbach alpha of .85, this factor represented entertainment and habit, and accounted for 4.2% of the variance.

Face validity for the Computer-Mediated Communication Motives Scale is established through the wording of the items. Content validity is supported by the way the scale has been constructed. Both Greenberg (1974) and A. Rubin (1981) ran pilot studies for the scale, and modified the items according to the outcome of the pretest. A. Rubin (1981) also reported construct validity when he compared rankings of his respondents' open-ended answers to their answers on the items of the scale and reported a Kendall's Coefficient of Concordance of .93. Finn and Gorr (1988) also supported the scale's construct validity when they predicted and reported that social compensation motivations (escape and companionship) were related to dissatisfaction with the social environment and a greater sense of loneliness and lack of social support (cited in R. Rubin et al., 1994).

Several studies also provided evidence of support for concurrent and criterion validity for the different motive dimensions. For example, A. Rubin and

Perse (1987a) reported an association between post-viewing discussions and watching soap operas for social utility motivations. A. Rubin and Perse (1987b) found that respondents who reported more distractions while watching the news scored high on motivations of passing time, and that there was a positive relationship between cognitive involvement with the news and watching for information seeking motivations.

Since the Computer-Mediated Communication Motives Scale has been developed and used in Europe and the United States, this study started out with the full a priori version of the scale to cover for the possibility of culture affecting Internet motivations differently in Egypt and the Arab world than it does in Europe and the United States. As a pilot study, the survey was administered to Arab undergraduate and graduate students of universities in Egypt (other than AUC) and the United States, as well as recent AUC graduates who had obtained their degrees within the previous two years. The participants were also asked to provide any qualitative comments or suggestions by e-mail to the researcher. A total of 50 respondents completed the pilot study. Based on their input, some minor changes were made to the survey instrument. Respondents indicated that the scale was "too long," so six items were taken out. Those were the items with the lowest mean scores in the pilot study. One item was added to the scale to account for a potential cultural difference in language, "To improve my English language." Based on respondents' suggestions, the wording of "To see what is out there" was changed to "To learn more about the outside world."

The next section of the questionnaire featured an Internet Affinity Scale, a modification of A. Rubin's (1981, 1983) Television Affinity Scale. A. Rubin used the scale on television viewing and reported an inter-item correlation of .44 and a reliability alpha coefficient of .75. A. Rubin and Perse (1987a) used a variation of the scale for soap opera viewing affinity and reported a Cronbach alpha of .80. A. Rubin (1985b), again on soap opera viewing affinity, reported a Cronbach alpha of .72. The Internet version of the scale used in this study is a 5-item Likert-type scale. It was used in Papacharissi and A. Rubin (2000) and has an established Cronbach alpha of .84.

This section of the questionnaire was followed by a four-part question measuring Arabic-language use of the Internet by respondents in terms of reading and writing. The final wording of those statements was also based on the input of the respondents in the pilot study. The question measured the respondents' use of Arabic and Arabic written in Latin characters (e.g., "*Sabah el Kheir*," which is Arabic for "good morning") in reading and writing on the Internet.

I also came up with a 4-item Cultural Identity Threat Scale that I included in the questionnaire to measure whether the Internet is perceived by respondents as an opportunity or a threat to their Arab cultural identity. The scale maintained face validity. It was pretested as part of the pilot study and had a Cronbach alpha reliability coefficient of .83.

Following Palmgreen and Rayburn (1985) and Papacharissi and Rubin (2000), satisfaction with Internet use was measured by a Likert-type scale item asking respondents how much they agreed or disagreed with the statement: "Overall, I am satisfied with the job the Internet does in providing me with the things I am looking for."

The final section of the questionnaire was concerned with demographics and psychographics. Demographic questions asked the respondents about their age, gender, year in school, and what school they belonged to within AUC. The questionnaire then inquired about how long respondents have been using the Internet (in years and months), as well as their level of operational skills (Internet skills, typing skills, English-language proficiency skills). Respondents were also asked to indicate whether they use a dial-up connection or a fast access connection at the place where they accessed the Internet the most.

Respondents' media ownership was assessed through a series of dichotomous questions about whether or not respondents owned a VCR, CD player, DVD player, mobile phone, computer or laptop, handheld organizer or PDA, or a satellite dish or cable subscription. Lastly, respondents were asked how the Internet has affected their patterns of social and other activities. They were asked, "Since you've started using the Internet, would you say the time you spend on these other activities has increased, decreased, or would you say it has not been affected?" Activities listed were: Watching TV (other than online), reading newspapers (other than online), reading magazines (other than online), making phone calls (other than online), studying, reading nonacademic books, listening to music (tapes or CDs), listening to the radio (other than online), going to the movie theater (cinema), watching movies on video or DVD, socializing with family (other than online), and socializing with friends (other than online).

I will now review the results and major findings of the study.

· 1 0 ·

THE USES AND GRATIFICATIONS OF THE INTERNET AMONG ARAB STUDENTS IN EGYPT: RESULTS AND FINDINGS

A total of 502 undergraduate and graduate students at AUC filled out the online survey. The data were analyzed using SPSS for Windows, Standard Version, Release 11.5.0.

Respondents' Demographics and Characteristics

Out of the 502 respondents, 58.5% were female ($n = 290$), and 41.5% ($n = 207$) were male.[1] This slight female overrepresentation matches the overall gender distribution of AUC students. As shown in Table 10.1, graduate students accounted for 28.5% of respondents ($n = 141$), freshmen for 20.5% ($n = 101$), sophomores for 17.0% ($n = 84$), juniors for 15.0% ($n = 75$), and seniors for 19.0% ($n = 94$). Respondents represented the three schools of AUC, with 21.5% ($n = 107$) from the School of Humanities and Social Sciences, 34.2% ($n = 170$) from the School of Sciences and Engineering, and

Table 10.1. Number and percentage of student respondents by gender and year in school

		Please indicate your gender		
		Male	Female	Total
Please indicate your year in school				
Freshman	*n*	43	58	101
	% of Total	8.7	11.7	20.5
Sophomore	*n*	37	47	84
	% of Total	7.5	9.5	17.0
Junior	*n*	27	47	75
	% of Total	5.5	9.5	15.0
Senior	*n*	32	62	94
	% of Total	6.5	12.6	19.0
Graduate	*n*	66	75	141
	% of Total	13.4	15.2	28.5
Total	*n*	205	289	494
	% of Total	41.5	58.5	100.0

37.8% (n = 188) from the School of Business, Economics, and Communications. The remaining 6.4% (n = 32) were undecided students.

Respondents were asked where they accessed the Internet the most. More than half (50.2%, n = 252) said they accessed it both "at home and at school." An additional 41% (n = 206) said they accessed the Internet from home. Only 6.4% (n = 32) said they accessed it from school, and 2.4% (n = 12) said they accessed it at other places.

Respondents ranged in age from 18 to 58 years. The mean age was 21.97, the median was 21, and the mode was 20. The majority of respondents (96.8%, n = 479) were between the ages of 18 and 30. More than two-thirds of respondents (71.1%, n = 352) were between the ages of 18 and 22, and only 2% were above 33 years of age (n = 10) (See Figure 10.1).

The mean length of time that respondents have been using the Internet was 5.71 years. Only 10.9% (n = 54) reported using the Internet for less than 4 years. Those using it for more than 7 years accounted for 15.4% (n = 78), while the majority of respondents (73.7%, n = 362) have been using the Internet for 4 to 7 years (Figure 10.2).

Respondents were asked about the time they spent on the Internet and its different applications in an average week. They reported spending an average

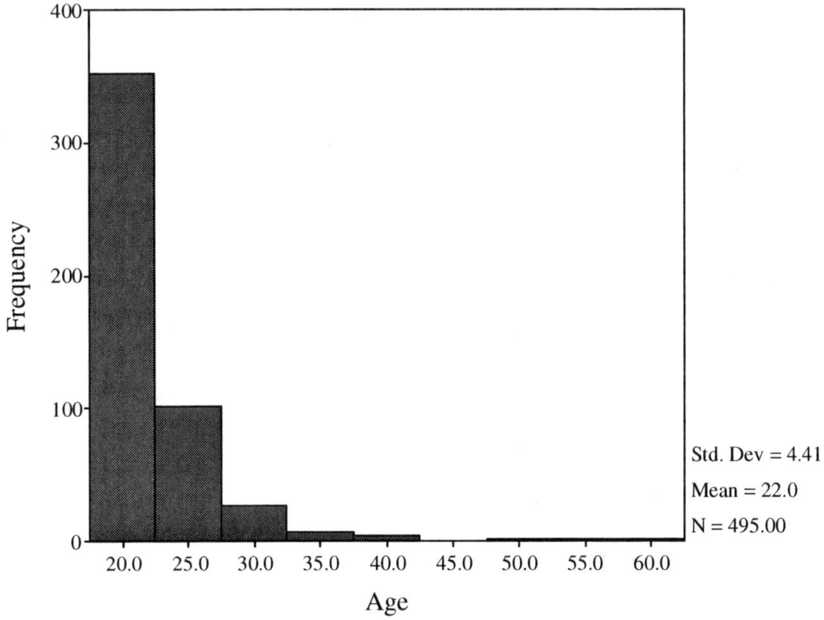

Figure 10.1. Frequency distribution of respondents' age.

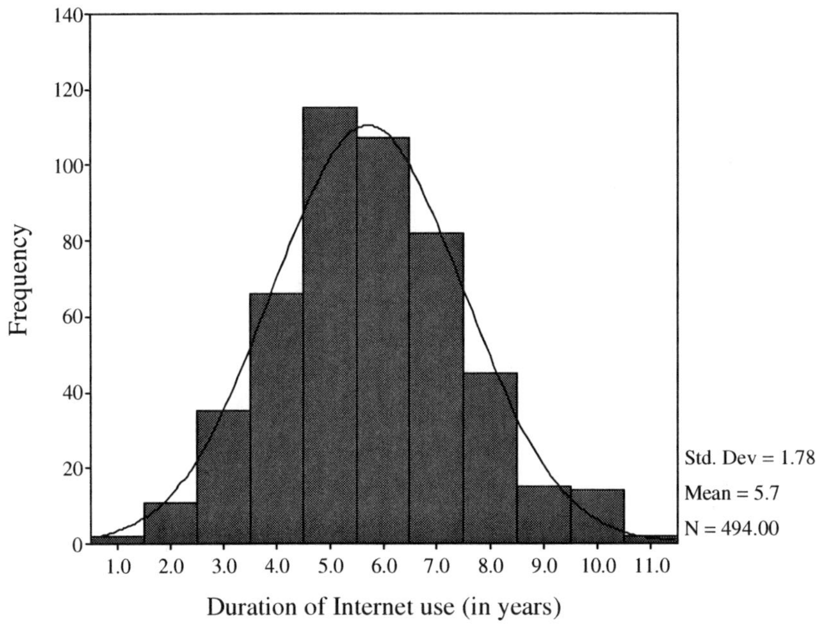

Figure 10.2. Duration of respondents' Internet use in years.

total time of 18.7 hours on the Internet, 4.5 hours on e-mail, 6.7 on the Web, 4.8 on chat/instant messaging/bulletin or message boards, and 3.0 hours on other activities. Table 10.2 shows the average time spent on the Internet classified by gender.

Students were asked about the type of Internet connection they used at the place where they accessed the Internet the most. Most (61.7%, n = 306) reported using a dial-up modem connection, while about one-third (32.3%, n = 160) reported using a fast access connection. The remaining 6% (n = 30) said they did not know.

The study attempted to gauge the level of Internet audience activity in terms of habitual versus instrumental users by asking two questions. One question asked, "When you use the Internet, would you say that most of the time you log on with a specific task in mind (such as searching for information, and/or checking e-mail), or do you just log on?" About four-fifths of respondents (78.4%, n = 393) indicated being instrumental users, saying they logged on with a specific task in mind, while one-fifth (21.6%, n = 108) indicated that they were habitual users, saying that they just logged on. The other question asked, "When you are online, how much attention would you say you pay to the task you are doing?" The majority of respondents (54.7%, n = 274) said "a lot," 42.5% (n = 213) said "some," while 2.4% (n = 12) said "very little," and 0.4% (n = 2) said "not at all." The mean score was 3.51 out of 4 (SD = 0.568).

Three questions inquired about respondents' self-assessment of their Internet operational skills. First, students were asked, "How well do you think

Table 10.2. Average time spent on the Internet by gender

Variable	Please indicate your gender					
	Male		Female		Total	
	Mean	n	Mean	n	Mean	n
Time spent on e-mail	4.26	205	4.76	289	4.55	494
Time spent on the Web	7.77	206	5.82	284	6.64	490
Time spent on chat/instant messaging/ bulletin or message boards	5.06	204	4.71	287	4.85	491
Time spent on other Internet activities	4.41	201	1.97	280	2.99	481
Total time spent on the Internet	21.20	207	17.01	290	18.76	497
Duration of Internet use (in years)	6.01	207	5.50	286	5.71	493

Note. All times are in hours during an average week.

you can use the Internet?" No students reported being beginner users, defined in the question as "I can barely do what I want." One-third of respondents (33.3%, n = 165) reported being intermediate users, defined as "I know my way around to the basics." And two-thirds (66.7%, n = 331) categorized themselves as expert users, defined as "I know my way very well on the Internet."

The second question that assessed Internet operational skills asked, "How fast would you say you can type?" The majority of respondents (61.8%, n = 306) reported being moderate typists, with 25–50 words per minute. Self-assessed slow typists, with under 25 words per minute, accounted for 14.3% (n = 71) of respondents, while fast typists, with more than 50 words per minute accounted for 23.8% (n = 118).

The third Internet operational skill was English-language proficiency. Most respondents (78.9%, n = 393) said they were advanced or "fluent" English speakers, defined as "I can write and converse on any topic, no problems." Only 1.2% (n = 6) said they were beginner or "okay" speakers, defined as "I can manage to write and have a conversation, but with some difficulty," and 19.9% (n = 99) said they were intermediate or "fair" speakers, defined as "I can write and converse on most topics, but I need to practice."

Students indicated a high level of satisfaction with the Internet. Only 2.2% (n = 11) indicated that they were not satisfied, while 45.1% (n = 223) said they were satisfied and another 45.1 (223) said they were very satisfied. The remaining 7.7% (n = 38) were neutral.

The study I conducted also investigated students' use of Arabic and/or Arabic in Latin characters. Four questions were designed to ask students about their use of the Arabic language on the Internet in terms of reading and writing, both in Arabic characters and in terms of Arabic dialogue written phonetically in English letters. Figures 10.3–10.6 display the use of the Arabic language by the student respondents.

Overall, Arabic was not very commonly used by respondents. About one-third (30.3%, n = 152) said they never read in Arabic on the Internet. Another 37.1% (n = 186) said they only used it 10% of the time (Figure 10.3). Respondents were even less likely to use Arabic to write on the Internet, with 77.1% (n = 387) saying they never use it, and another 13.9% (n = 70) saying they used it only 10% of the time (Figure 10.4).

Respondents were slightly more likely to use Arabic words spelled phonetically in English letters and characters. For example, "*sabah el kheir*," which is Arabic for "good morning," would be written in Latin characters to sound like the Arabic words. Figures 10.5 and 10.6 show a distribution of respondents' use of such dialogue both in terms of reading and writing.

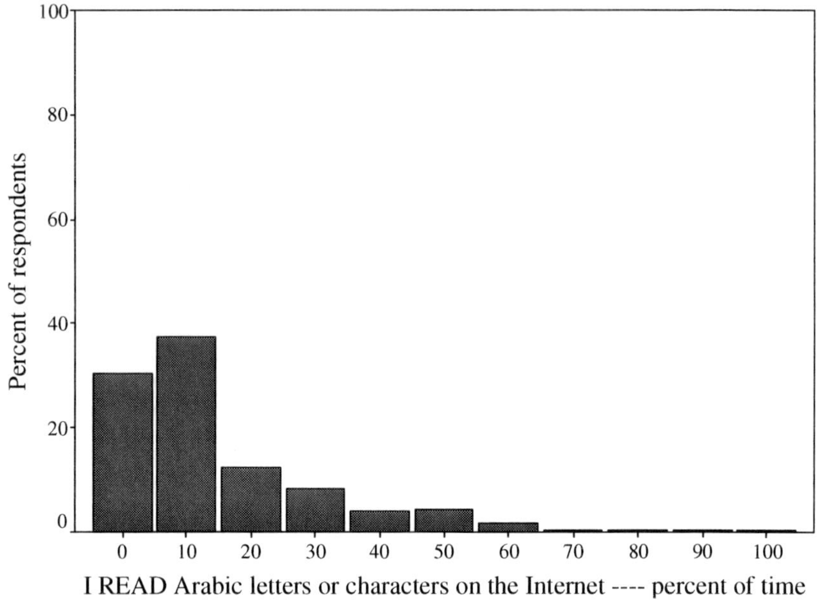

Figure 10.3. I READ Arabic letters or characters on the Internet—percent of the time.

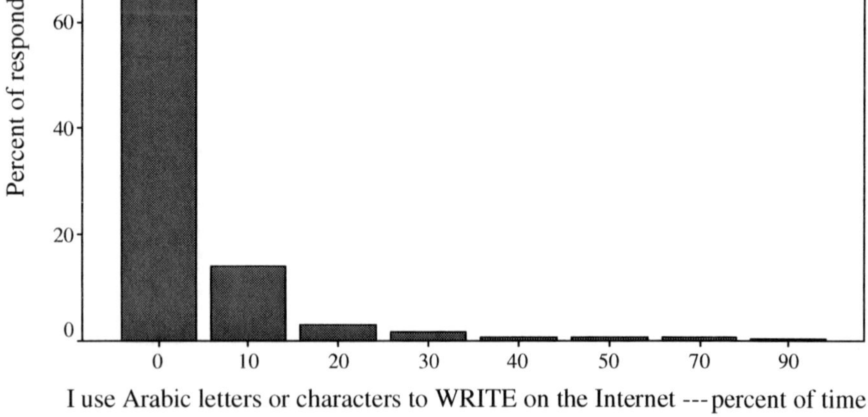

Figure 10.4. I use Arabic letters or characters to WRITE on the Internet—percent of the time.

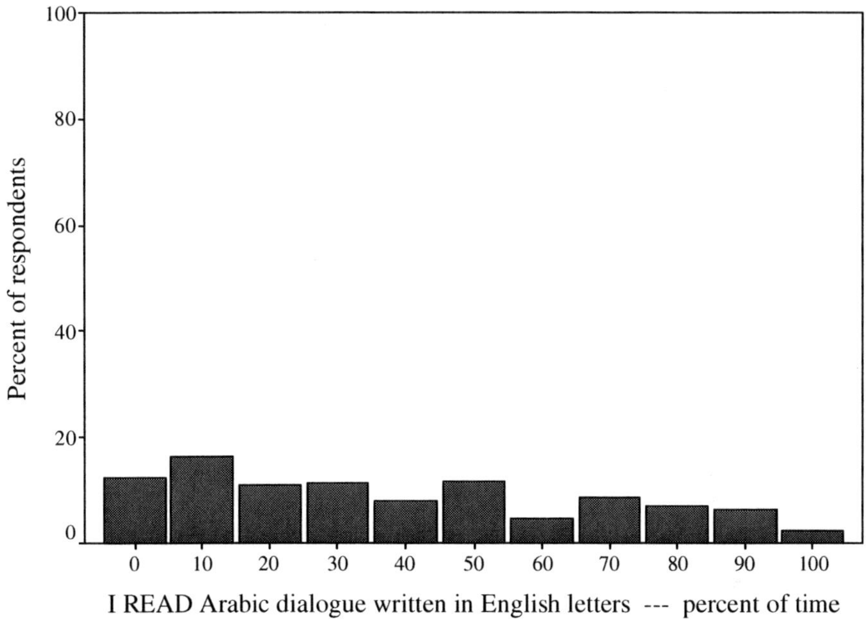

Figure 10.5. I READ Arabic dialogue written in English letters or characters on the Internet (e.g., Sabah el kheir)—percent of the time.

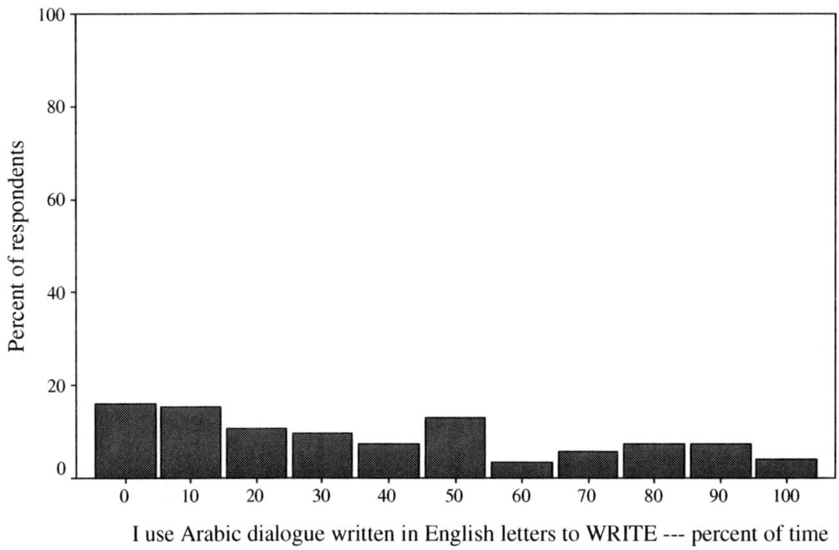

Figure 10.6. I use Arabic dialogue written in English letters or characters on the Internet (e.g., Sabah el kheir) to WRITE—percent of the time.

Table 10.3. Means and standard deviations of variables on the Cultural Identity Threat Scale

Variable	Mean	Standard Deviation
The Internet poses a threat to the Arab identity and culture.*	3.67	1.116
The Internet could help Arabs learn about the rest of the world.	4.48	0.616
The Internet could help Arabs express their identity, views, and culture to the rest of the world.	4.50	0.673
The Internet presents an opportunity for the Arab world.	4.31	0.804
Scale summated mean	4.24	0.80

Note. * Item is reverse coded.

Table 10.4. Patterns of media ownership among respondents

Media ownership: Do you have a . . .	Percentage of "Yes" responses	Frequency of "Yes" responses (n)
Videocassette recorder (VCR)	91.7	454
CD player (regular or portable)	91.3	452
DVD player (regular or portable)	47.6	234
Mobile phone	97.6	483
Home computer or laptop	98.8	490
Electronic handheld organizer (PDA)	26.4	130
Satellite dish or cable access	83.5	415

I devised a 4-item Likert-type Cultural Identity Threat Scale that asked respondents whether they thought the Internet could help Arabs in terms of learning more about the world and spreading their own culture and traditions, or whether they saw it as a threat to their cultural heritage. The majority of respondents (61.5%, n = 304) did not view the Internet as posing a threat to Arab identity or culture. Only 18.6% (n = 92) agreed that the Internet was a threat. An overwhelming 95.1% (n = 472) agreed that "The Internet could help Arabs learn about the rest of the world." Similarly, 93.2% (n = 462) agreed that "The Internet could help Arabs express their identity, views, and culture to the rest of the world." And 84.3% (n = 418) agreed that "The Internet presents an opportunity for the Arab world." Table 10.3 shows the means and standard deviations for the variables on the Cultural Identity Threat Scale. The scale had a summated mean of 4.24 (SD = 0.80), and a standardized item alpha of .71.

Respondents were asked a series of dichotomous questions to assess their patterns of media ownership. The results of these questions are shown in Table 10.4.

Research Questions and Hypotheses Testing

RQ1: *How do Arab English-speaking university students in Egypt use the Internet?*

Principal components factor analysis on the Computer-Mediated Communication Motives Scale was conducted using Varimax orthogonal rotation. The initial model extracted 10 factors that met the Kaiser criterion (had an eigenvalue of at least 1.0). Those explained 61.95% of the variance. However, given that Varimax rotation is likely to produce a magnified number of factors, a Scree plot was also employed, as well as a sense of comprehensibility and interpretability (Gorsuch, 1983; Harman, 1976; Kaiser, 1960, 1970, 1974; Kim & Mueller, 1978a, 1978b; Kruskal & Wish, 1978; Norusis, 2002; Williams, 1992). A conservative standard was employed, whereby factors retained were those that met the Kaiser criterion *and* had a minimum of three items loading with .60 or higher on the factor and no higher than .40 on any other factor. The model resulted in five interpretable factors comprised of 22 variables and explaining 47.27% of the variance. Items that did not load on any retained factor were eliminated from further analysis. Table 10.5 displays the factor analysis for Internet motives.

The first factor, *Personal Utility*, accounted for 11.2% of the variance after rotation. The factor had seven items loading on it, combining items from the a priori categories of escape, pass time, and inclusion. It had a standardized Cronbach alpha of .84 and an eigenvalue of 8.24.

Social Interaction was the second factor, accounting for 9.6% of the variance after rotation. It had five items of social inclusion and expressive needs. The factor had an eigenvalue of 4.47 and a standardized alpha of .81.

The third factor had four items, mostly pertaining to *Entertainment*. It had an eigenvalue of 2.88 and a standardized alpha of .79. It accounted for 7.8% of the variance after rotation.

Table 10.5. Factor analysis for Internet motives

I use the Internet . . .	Component				
	1	2	3	4	5
Factor 1: Personal Utility					
Because it passes the time away, particularly when I'm bored	**.725**	.003	.280	−.067	.023
When I have nothing better to do	**.674**	.142	.238	−.073	.040
Because it gives me something to do to occupy my time	**.664**	.071	.348	.033	.039
Because it makes me feel less lonely	**.651**	.196	.036	.061	−.015
So I can get away from what I'm doing	**.651**	.009	.089	.094	−.088
So I can forget about school, work, or other things	**.650**	.091	.227	−.001	−.162
When there's no one else to talk to or to be with	**.604**	.094	.149	.016	−.084
Factor 2: Social Interaction					
To belong to a group with the same interests as mine	.063	**.762**	.005	.100	.004
Because I enjoy answering other people's questions	.111	**.762**	.085	.097	−.012
To participate in discussions	.084	**.683**	.189	.202	.031
To help others	.084	**.620**	−.093	.157	.189
To show others encouragement	.087	**.603**	.025	.353	.013

Factor 3: Entertainment					
Because I just like to use it	.167	.008	**.759**	.039	.030
Because it's entertaining	.244	.093	**.731**	.015	.210
Because it's fun just to play around and check things out	.304	.049	**.684**	.083	−.102
Because it's enjoyable	.229	.157	**.633**	.103	.367
Factor 4: Surveillance					
To learn more about the outside world	.008	.056	.042	**.692**	.368
To keep up with current issues and events	.009	.071	−.094	**.675**	.272
To give my input	.026	.298	.104	**.627**	.051
Factor 5: Information					
To search for information	−.083	.042	−.013	.135	**.742**
Because it provides a new and interesting way to do research	−.087	.133	.158	.199	**.688**
Because it's easy to get information	−.095	−.062	−.026	.177	**.655**

Note. Factor 1 (Personal Utility) had an eigenvalue of 8.24, Factor 2 (Social Interaction) had an eigenvalue of 4.47, Factor 3 (Entertainment) had an eigenvalue of 2.88, Factor 4 (Surveillance) had an eigenvalue of 1.75, and Factor 5 (Information) had an eigenvalue of 1.58.

Surveillance was the theme of the fourth factor, which included three items. The factor explained 6.4% of the variance after rotation. It had an eigenvalue of 1.75 and a standardized alpha of .67.

The fifth factor also had three items, pertaining to *Information*. This factor had an eigenvalue of 1.58 and a standardized alpha of .71. It accounted for 6.3% of the variance after rotation.

The fifth factor, information, had by far the highest mean ($M = 4.55$, $SD = 0.50$). Surveillance followed with a mean of 3.94 ($SD = 0.70$). Entertainment had a mean of 3.71 ($SD = 0.74$). Personal utility had a mean of 3.12 ($SD = 0.85$), and social interaction had the lowest mean of 2.94 ($SD = 0.76$). Table 10.6 shows the mean scores and standard deviations for Internet motives.

Pearson r correlations were also computed among the different motives. The strongest significant correlations were between personal utility and entertainment ($r = .53$, $p <.001$), social interaction and surveillance ($r = .44$, $p < .001$), and information and surveillance ($r = .42$, $p < .001$). Table 10.7 shows the Pearson r correlations among all motives.

Therefore, in response to the research question, Arab English-speaking university students in Egypt use the Internet primarily for information and surveillance. They also use it for entertainment, to escape, pass time, or seek companionship, and for social interaction.

H1: *There is a significant difference in Internet use motives between males and females among Arab English-speaking university students in Egypt.*

H2: *There is a significant difference in Internet exposure in terms of time spent on the Internet between males and females among Arab English-speaking university students in Egypt.*

The next variable I examined was gender and how it affects Internet use motives among respondents. To test the first hypothesis, a series of independent sample t-tests was conducted to examine how gender affects Internet use motives among respondents. Significant differences were found in motives of social interaction and information seeking. Contrary to popular stereotypes, males were more likely than females to use the Internet for social interaction, $t(491) = 3.48$, $p = .001$, while females were more likely than males to use it for information seeking, $t(495) = -2.61$, $p = .009$. Some other differences were present although they did not prove statistically significant. Males were slightly more likely to use the Internet for escape, to pass time, and for companionship, as well as for surveillance, and females were slightly more likely to

Table 10.6. Means and standard deviations for Internet motives

I use the Internet . . .	Mean	SD
Factor 1: Personal Utility	**3.12**	**0.847**
Because it passes the time away, particularly when I'm bored	3.68	1.199
When I have nothing better to do	3.24	1.161
Because it gives me something to do to occupy my time	3.23	1.132
Because it makes me feel less lonely	2.92	1.238
So I can get away from what I'm doing	2.95	1.178
So I can forget about school, work, or other things	2.61	1.118
When there's no one else to talk to or to be with	3.24	1.231
Factor 2: Social Interaction	**2.94**	**0.760**
To belong to a group with the same interests as mine	2.77	1.032
Because I enjoy answering other people's questions	2.74	1.021
To participate in discussions	3.07	1.054
To help others	3.08	0.907
To show others encouragement	3.03	0.979
Factor 3: Entertainment	**3.71**	**0.743**
Because I just like to use it	3.46	1.041
Because it's entertaining	3.91	0.818
Because it's fun just to play around and check things out	3.48	1.062
Because it's enjoyable	3.96	0.856
Factor 4: Surveillance	**3.94**	**0.698**
To learn more about the outside world	4.19	0.840
To keep up with current issues and events	4.14	0.841
To give my input	3.46	1.004
Factor 5: Information	**4.55**	**0.496**
To search for information	4.64	0.579
Because it provides a new and interesting way to do research	4.35	0.747
Because it's easy to get information	4.64	0.546

Note. Items were rated on a 5-point Likert-type scale. Responses ranged from "Strongly disagree" (1) to "Strongly agree" (5).

Table 10.7. Pearson *r* correlations among Internet motives

	Personal utility	Social interaction	Entertainment	Surveillance	Information
Personal utility	1				
Social interaction	.266***	1			
Entertainment	.533***	.192***	1		
Surveillance	.079	.443***	.188***	1	
Information	−.079	.164***	.140**	.422***	1

Note. ** Correlation is significant at $p < .01$ level (2-tailed); *** Correlation is significant at $p < .001$ level (2-tailed).

use it for entertainment. Based on the overall analysis, gender was shown to be a significant factor affecting some Internet use motives. ***H1*** was therefore accepted.

To test for ***H2***, a series of *t*-tests was conducted to look for mean differences between males and females in terms of Internet exposure. Gender proved to be a significant factor [$t(495) = 2.81$, $p = .005$] in the overall time spent on the Internet during an average week. Males spent significantly more time on the Internet (M = 21.20, *SD* = 12.70) than females (M = 17.01, *SD* = 15.38). Furthermore, males spent significantly more time on the Web (M = 7.76, *SD* = 7.80) than females (M = 5.82, *SD* = 5.46), with $t(488) = 3.24$, $p = .001$. Males also spent significantly more time on other Internet activities (M = 4.41, *SD* = 6.79) than females (M = 1.97, *SD* = 3.99), with $t(479) = 4.93$, $p < .001$. Females spent more time on e-mail than males (as shown earlier in Table 10.2), but the mean differences were not statistically significant. Based on the overall analysis, gender was a significant factor in the amount of Internet exposure for these respondents. **H2** was therefore accepted.

RQ2: *Is school year a significant factor in how Arab English-speaking university students use the Internet?*

To examine the year in school variable, a series of one-way analysis of variance (ANOVA) tests was carried out between year in school and the five Internet use motives. The only motive that proved significant was personal utility, with $F(4, 488) = 3.44$, $p = .009$. To determine which mean scores were significantly different, *t*-tests between all possible combinations of groups in relation to the personal utility motive were then conducted. There was a significant difference between freshmen and seniors [$t(192) = 2.51$, $p = .013$], and freshmen and graduate students [$t(239) = 2.57$, $p = .011$]. Mean scores showed that freshmen were significantly more likely to use the Internet for personal utility reasons than seniors or graduate students.

H3: *There is a significant positive correlation between Internet affinity and Internet use motives among Arab English-speaking university students.*

Next I examined Internet affinity. Affinity was measured by a 5-item Likert type scale. Response choices ranged from strongly disagree (1) to strongly agree (5). The summated mean for the scale was 3.43 (*SD* = 0.84), indicating that respondents found the Internet to be relatively important in their lives. The scale had a reliability Cronbach alpha of .81.

To test the hypothesis, Pearson *r* correlations were conducted, first between Internet affinity and the summated mean for Internet use motives, and then between Internet affinity and each of the motives. There was a statistically significant, moderate, positive correlation between Internet affinity and the summated mean for Internet use motives ($r = .45$, $p < .001$). Furthermore, Internet affinity was correlated to all individual factor mean scores for Internet use motives. The strongest correlation ($r = .36$, $p < .001$) was with entertainment. Internet affinity was also correlated to personal utility ($r = .27$), social interaction ($r = .28$), surveillance ($r = .29$), and information ($r = .22$), all at $p < .001$. Internet affinity also correlated significantly with Internet exposure in terms of total time spent on the Internet in an average week ($r = .23$, $p < .001$), and with Internet satisfaction ($r = .22$, $p < .001$). Hypothesis 3 was therefore accepted.

H4: *There is a significant positive correlation between Internet satisfaction and Internet use motives among Arab English-speaking university students.*

The mean score for satisfaction with the Internet was 4.33 out of 5 ($SD = 0.71$), indicating that respondents were highly satisfied with their Internet experiences. To test the hypothesis, a series of Pearson r correlations was conducted between Internet satisfaction and Internet use motives. There was a weak but statistically significant positive correlation between Internet satisfaction and the summated mean for Internet motives ($r = .24$, $p < .001$). Internet satisfaction also correlated significantly to Internet use motives of entertainment ($r = .19$), surveillance ($r = .18$), and information ($r = .33$), all at $p < .001$. As reported above, Internet satisfaction correlated significantly with Internet affinity ($r = .22$, $p < .001$). However, it did not correlate significantly with Internet exposure. Based on this analysis, hypothesis 4 was accepted.

H5: *There is a significant difference between high and low levels of operational skills necessary for Internet use (typing speed, English-language proficiency, Internet use proficiency) and Internet use motives among Arab English-speaking university students in Egypt.*

H6: *There is a significant difference between high and low levels of operational skills necessary for Internet use (typing speed, English-language proficiency, Internet use proficiency) in terms of time spent on the Internet among Arab English-speaking university students in Egypt.*

These two hypotheses concern the operational skills needed for Internet use, measured in this study through three different ordinal-level variables: typing speed, English-language proficiency, and Internet use proficiency. In terms of typing speed, most respondents (61.8%, $n = 306$) said they typed at a moderate speed of between 25 and 50 words per minute. Fast typists (over 50 words per minute) constituted 23.8% ($n = 118$) of the student respondents, and slow typists (under 25 words per minute) were 14.3% ($n = 71$).

Most students (78.9%, $n = 393$) reported a high level of English-language proficiency. Almost one-fifth (19.9%, $n = 99$) reported a moderate proficiency level, and only 1.2%, ($n = 6$) reported a low proficiency level of English language.

As for Internet use proficiency, no respondents reported a low level. One-third of respondents reported being intermediate users (33.3%, $n = 165$), and two-thirds (66.7%, $n = 331$) reported being expert users.

To test for hypothesis 5, one-factor ANOVAs were conducted to test for differences between each of the operational skills and the Internet use motives. Neither typing speed nor English-language proficiency showed significant effects on any of the motives. Internet use proficiency, however, did. A *t*-test (since there were only two categories on the Internet use efficiency variable) showed significant differences between intermediate and expert users in terms of their reasons for using the Internet on the entertainment [$t(491) = -1.97$, $p = .049$], surveillance [$t(493) = -2.31$, $p = .021$], and information [$t(494) = -2.90$, $p = .004$] motives. Expert users were more likely to use the Internet for all five motives. However, differences on the personal utility and social interaction motives were not statistically significant. Internet use proficiency therefore appeared to be a significant factor in how these students used the Internet. Hypothesis 5 was therefore accepted.

To test for hypothesis 6, one-factor ANOVAs and *t*-tests were conducted to test for differences between each of the operational skills and total time spent on the Internet during an average week (exposure). Typing speed and Internet use proficiency both proved significant. Fast typists ($M = 21.77$, $SD = 17.75$) spent significantly more time on the Internet during an average week [$t(187) = -2.46$, $p = .015$] than slow typists ($M = 15.51$, $SD = 15.48$). Expert Internet users ($M = 20.53$, $SD = 17.37$) also spent significantly more time on the Internet during an average week [$t(494) = -3.49$, $p = .001$] than intermediate users ($M = 15.10$, $SD = 14.01$). This hypothesis was therefore accepted.

H7: *There is a positive correlation between Internet exposure and Internet use motives among Arab English-speaking university students.*

Internet exposure in terms of the average time spent by respondents on the Internet per week (Table 10.2) was correlated against Internet use motives. The results indicated that Internet exposure was significantly correlated to the summated mean for Internet use ($r = .27, p < .001$). Internet exposure was also significantly correlated to Internet use motives of personal utility ($r = .18$), social interaction ($r = .24$), and entertainment ($r = .23$), all at $p < .001$, and to surveillance ($r = .13$) at $p < .01$. Hypothesis 7 was therefore accepted.

RQ3: *Is type of Internet connection (dial-up modem versus fast access) a significant factor in how Arab English-speaking university students use the Internet?*

RQ4: *Is type of Internet connection (dial-up modem versus fast access) a significant factor in the amount of time Arab English-speaking university students spend on the Internet?*

The next variable examined was type of Internet connection. To address **RQ3**, a series of *t*-tests was conducted to look at mean differences in Internet use motives between users of dial-up modem and users of fast access Internet connections. There were no significant differences between means of Internet use motives for dial-up modem and fast access Internet connection users. Therefore, type of Internet connection did not appear to be a significant factor in how respondents used the Internet.

To test for **RQ4**, another series of *t*-tests was conducted to look at mean differences in Internet exposure in terms of time spent on the Internet in an average week between users of dial-up modem and users of fast access Internet connections. The *t*-tests showed a significant difference in overall Internet exposure between means of dial-up modem and fast access Internet connection users [$t(464) = -3.171, p = .002$]. Those who had fast access to the Internet were significantly more likely to spend more time ($M = 22.63$, $SD = 19.26$) on the Internet in an average week than those who accessed the network through a dial-up modem ($M = 17.50$, $SD = 15.00$). There was also a significant difference between the two groups in terms of time spent on e-mail [$t(461) = -3.498$, $p = .001$], and time spent on the Web [$t(457) = -3.571, p < .001$]. Fast access users also spent more time during an average week on chat, instant messaging, and discussion boards, as well as on "other" Internet functions. These differences however were not statistically significant. It was concluded therefore that the type of Internet connection was a significant factor in Internet exposure or the amount of time respondents spent on the Internet.

RQ5: *How does Internet use affect use of other media among Arab university students in Egypt?*

RQ6: *How does Internet use affect social activities among Arab university students in Egypt?*

To gain insight into these two research questions, respondents were asked to indicate whether their time spent on a list of media and social activities has (1) decreased, (2) not been affected, or (3) increased since they started using the Internet. A summary of their answers is presented in Table 10.8.

As the table indicates, the majority of respondents believe their media and social habits and behaviors have not been affected by their use of the Internet. Over 70% of respondents said Internet use has not affected their time to socialize with family, socialize with friends, or go to the movie theater. Over 60% believe that Internet use has not affected their reading newspapers, reading magazines, listening to radio, making phone calls, or watching movies. The medium that seemed to have been affected the most was television, with 45.3% of respondents saying they now watch less television compared to earlier days before they began using the Internet. Some activities, such as reading, have been negatively affected though. For example, 33.1% of respondents reported less time reading newspapers, less time reading magazines (30.4%), and less time reading nonacademic books (29%). Some respondents (28.3%) also reported less time studying. Socializing with family has decreased with 21.9% of respondents, and making phone calls with 26.3%. Listening to music seemed to be the activity that had the most increase, with 35.8% of respondents reporting that they now listen to more music on tapes or CDs.

RQ7: *What variables predict Internet exposure among Arab English-speaking university students in Egypt?*

RQ8: *What variables predict Internet use among Arab English-speaking university students in Egypt?*

The final research questions examined the variables that predicted Internet exposure and Internet use among respondents. To answer these questions, a series of forward multiple regression analysis models was used. A probability criterion of *F*-to-enter $< .05$ was employed. To test for any potential problems of normality violations or multicollinearity, "collinearity diagnostics" were requested with the SPSS output. As suggested by J. Cohen and P. Cohen (1983) and Harris (2001), the criteria employed for detecting problems with multicollinearity were a tolerance of less than .20 and a variance-inflation factor (VIF) of more than 4. The lowest tolerance found was .65, and the highest VIF was 1.55. Therefore, no such problems were found and the regression analyses were deemed appropriate.

Significant predictors to Internet exposure in terms of total time spent on the Internet in an average week were social interaction motive ($\beta = .196$,

Table 10.8. Self-reported changes in respondents' media and social behavior since they started using the Internet

Media and social behaviors	Time spent has decreased (%)	Time spent has not been affected (%)	Time spent has increased (%)
Watching TV (other than online)	45.3	52.3	2.4
Reading newspapers (other than online)	33.1	61.7	5.3
Reading magazines (other than online)	30.4	64.7	4.9
Reading nonacademic books	29.0	57.2	13.8
Listening to the radio (other than online)	32.2	62.5	5.3
Making phone calls (other than online)	26.3	67.6	6.1
Watching movies on video or DVD	18.8	62.9	18.4
Going to the movie theater (cinema)	15.1	75.8	9.2
Listening to music (tapes or CDs)	19.8	44.3	35.8
Studying	28.3	56.8	14.9
Socializing with family (other than online)	21.9	72.8	5.3
Socializing with friends (other than online)	15.4	70.6	14.0

$t = 4.244, p < .001$), entertainment motive ($\beta = .127, t = 2.656, p = .008$), type of Internet connection ($\beta = -.137, t = -3.085, p = .002$) (implying that those with fast access connections were more likely to spend time on the Internet), and affinity ($\beta = .115, t = 2.383, p = .018$). The model was statistically significant in predicting Internet exposure, $R = .35$, $R^2 = .12$, *Adjusted* $R^2 = .114$, $F(4, 450) = 15.552, p < .001$.

To answer the final research question, an attempt was made to predict specific patterns of Internet use in terms of different Internet functions. The first regression model aimed at predicting e-mail use by respondents (in terms of time spent on e-mail). Significant predictors were type of Internet connection ($\beta = -.155, t = -3.351, p = .001$), surveillance motive ($\beta = .112, t = 2.418, p = .016$), and age ($\beta = .103, t = 2.220, p = .027$). The model significantly predicted amount of time using e-mail, $R = .233$, $R^2 = .054$, *Adjusted* $R^2 = .048$, $F(3, 448) = 8.560, p < .001$.

Next, Web use was predicted (considered in terms of time spent by respondents in an average week). The regression model had six significant predictors. Those were: Entertainment motive ($\beta = .149, t = 3.061, p = .002$), social interaction motive ($\beta = .116, t = 2.465, p = .014$), type of Internet connection ($\beta = -.132, t = -2.938, p = .003$) (again implying that those with fast access connections were more likely to spend time using the Web), affinity ($\beta = .117, t = 2.405, p = .017$), gender ($\beta = .109, t = 2.392, p = .017$) (implying that males are likely to spend more time on the Web than females), and age ($\beta = .092, t = 2.036, p = .042$). The model was statistically significant in predicting time spent on the Web, $R = .355$, $R^2 = .126$, *Adjusted* $R^2 = .114$, $F(6, 442) = 10.653, p < .001$.

Finally, predictors for time spent on chat, instant messaging, and bulletin or discussion boards were examined. This regression model had five significant predictors. Those were: Personal utility motive ($\beta = .107, t = 1.939, p = .050$), age ($\beta = -.158, t = -3.537, p < .001$), social interaction motive ($\beta = .172, t = 3.640, p < .001$), information motive ($\beta = -.131, t = -2.800, p = .005$), and entertainment motive ($\beta = .141, t = 2.644, p = .008$). The model was statistically significant in predicting time spent on the chat, instant messaging, and bulletin or discussion boards, $R = .365$, $R^2 = .133$, *Adjusted* $R^2 = .123$, $F(5, 444) = 13.620, p < .001$.

This chapter provided a detailed presentation of the results of this study, answering the research questions and testing the hypotheses posed earlier. We now move to a discussion of the interpretations and the implications of these findings.

·11·

THE USES AND GRATIFICATIONS OF THE INTERNET AMONG ARAB STUDENTS IN EGYPT: DISCUSSION

The research study reviewed in this book assessed the uses and gratifications of the Internet among Arab English-language speaking university students in Egypt. I started by introducing a model of the historical development of communication technologies in the Arab world, and how each major development has affected the region. I then looked at telecommunications technology in the Arab world, gave a brief history of the Internet, and assessed the current status and future trends of the Internet in Arab countries. I looked at the role of Egypt historically as a leader in providing media content in the Arab world, and its potential role as a trend-setting power in providing Internet content as well. I also discussed the online free flow of information and how Arab governments are dealing (or not dealing) with the Internet, as well as how Islam as a religion is represented on the Internet and how can the network impact the religion or vice versa. I moved on to introduce the theoretical framework of uses and gratifications, and why it is well suited to studying the Internet at this stage of its development as a medium, and I reviewed the important literature on uses and gratifications, particularly as it related to the Internet and new media technologies. I also reviewed the studies relevant to the Internet in Egypt and the Arab world.

In this chapter, I will present you with a discussion of the major findings of the study and what it means to Egypt and the Arab world in simple terms, so you can get the important findings of it even if you are not well versed in research and statistical jargon. This study used an online survey administered to 502 students. The number of females was slightly more than males (58.5%), which actually matched the overall gender distribution of AUC students. Students were from all school years and from the three schools of AUC. (A more detailed examination of the sample demographics and psychographics was presented in the preceding chapter.)

More than half of the respondents indicated that they accessed the Internet both "at home and at school." An additional 41% said they accessed the Internet from home, while a minority said they accessed it only from school or other places such as Internet cafes. More than two-thirds of respondents were between the ages of 18 and 22. The average time they reported spending on the Internet in any given week was 18.7 hours, and the average time they have been using the Internet was 5.7 years. About two-thirds of the students said they had a dial-up modem connection, while about one-third reported using a fast access connection.

One-third of the students reported being intermediate Internet users, saying they knew their way around to the basics. Two-thirds categorized themselves as expert users. About 60% of the students reported being moderate typists (about 25–50 words per minute). Most students (78.9%) said they were advanced or "fluent" English speakers. A great majority of the students (over 90%) reported being either satisfied or very satisfied with the Internet.

One thing I investigated was students' use of Arabic and/or Arabic in Latin characters. Overall, Arabic was not commonly used by respondents. About 67% said they never read in Arabic on the Internet or they only use the language 10% of the time. More than 77% said they never use Arabic to write on the Internet. Students sometimes (although not too often) use Arabic words spelled in English letters (e.g., "*hello*," would be written as "*ahlan wa sahlan*,") so the Arabic words would be phonetically written in Latin characters.

I also investigated whether those students thought the Internet was a cultural threat or a cultural opportunity for Arabs, whether it could be used by Arabs to learn more about the world and spread their own culture and traditions, or whether it was a threat to their cultural heritage. The majority of students did not view the Internet as a threat to Arab culture. Actually, more than 95% thought of the Internet as a medium that "could help Arabs learn about the rest of the world." Also more than 93% agreed that "The Internet could help Arabs express their identity, views, and culture to the rest of the world," and 84% agreed that "The Internet presents an opportunity for the Arab world."

One of the important questions that this study answered was how students used the Internet. The study factor analyzed the main motives of Internet use among these students, and investigated several demographic and psychographic variables in relation to these motives. The variables I examined included gender, year in school, type of Internet connectivity, level of skills needed, Internet exposure, Internet affinity, and Internet satisfaction. I also explored patterns of other media and technology ownership among Internet users, as well as the potential effects, if any, of Internet use on patterns of social relationships and media consumption. Finally, I tried to assess the variables that predicted Internet use and Internet exposure among Arab students in Egypt.

The study indicated five primary motives for using the Internet among respondents. The most important use of the Internet for these students was information seeking (searching for information and doing research), followed by surveillance (learning more about the outside world and keeping up to date with current events). This indicated a more instrumental use of the Internet on the part of respondents, emphasizing that these are active audiences who know what they want out of the medium. The findings about the motives concur with past research of different Internet user populations (see, for example, Charney & Greenberg, 2002; Ebersole, 2000; Papacharissi & A. Rubin, 2000; Perse & Ferguson 2000). The findings also indicate that student Internet users in the Arab world are aware of the value of the Internet as an alternative, rich, source of information and news, and are keen on making use of this important function of the network.

Other motives of Internet use that emerged from the factor analysis were entertainment, personal utility, and social interaction. The entertainment factor included "playing around and checking things out." Personal utility included passing time, getting over boredom, getting over loneliness, and escape from work or life pressures. Social interaction included participating in discussions, helping others, and getting together with people who share the same interests as the user. These factors underline the interactive nature of the Internet and its potential to act as a functional alternative to other media or face-to-face communication, especially, as earlier research has concluded (see, for example, Papacharissi & A. Rubin, 2000; A. Rubin & R. Rubin, 1985), among users who are not very satisfied with their social lives or interpersonal communication skills. From a cultural perspective, it gives way to the possibility of potential across-borders friendships and acquaintances that could help teach members of different cultures about each other, potentially bridging a cultural gap. This is true both within the Arab world, and between the Arab world and the Western world.

An interesting insight that these results yield is that Internet use is motivated by similar reasons across both the Western and the Arab cultures. Information seeking has been rated as the most salient Internet use motive in this study as well as in earlier research done on U.S. populations (for example, Charney & Greenberg, 2002; Papacharissi & A. Rubin, 2000; Perse & Ferguson 2000). Some differences, however, might reflect on the different cultures. Social interaction, for example, scored the lowest means across the Arab sample. This might be due to the more conservative nature of the Arab culture when compared to its Western counterpart. Meeting "strangers" through media is also very new to the Arab world, where other more traditional media such as the telephone or print media are not used for purposes of meeting new people.

As hypothesized for this study within the uses and gratifications approach, gender was a significant factor. Males spent significantly more time on their overall Internet activities and Web endeavors than females. Males were also more likely to use the Internet for social interaction than females, and females were more likely to use it for information seeking purposes. Culture may also explain this finding, since males in the Arab world are always expected to take the initiative in meeting a member of the opposite sex, while females are almost never expected to do so. Males are therefore more accustomed to "starting a conversation with a stranger" than females, and the Internet provides a new, exciting venue for them to do this. And since online social interaction is time consuming, this would be a factor as to why males are likely to spend more time than females on the Internet in general and on the Web in particular.

Year in school was a significant factor on the personal utility motive. Seniors and graduate students were significantly less likely than freshmen to use the Internet for personal utility, which included a priori criteria of escape, passing time, and habit. This is likely to be caused by the increase of scholarly demands on students as they progress in their academic lines of study. This concurs with previous research (Atkin, Jeffres, & Neuendorf, 1998; Mesbah, 1991), which reported that higher educational levels go with more intentional Internet adoption practices or media use.

This study also investigated Internet affinity or importance to the user and Internet satisfaction by the user. As mentioned before, over 90% of the student sample reported being either satisfied or very satisfied with the Internet. As for affinity, respondents also indicated that the Internet has become important in their lives (scoring 3.43 out of 5). The results showed a positive, statistically significant correlation between affinity and all Internet use motives, which means that the more important the Internet is to the user, the more they relate to all five Internet use motives. There was also a positive, statistically significant relationship between Internet satisfaction and the motives of entertainment, surveillance, and information. This means that the more users are satisfied by the Internet, the more they relate to these three motives.

I also investigated the relationship between the three concepts of Internet exposure, affinity, and satisfaction. People mostly assume that if you get much exposure to something, then it must be important to you and you must be satisfied with it. The results showed that only the first half of this statement is correct. There was a positive, statistically significant correlation between exposure and affinity, and also between affinity and satisfaction, which meant that the more you are satisfied with the Internet, the more it becomes important to you. However, there was no significant relationship between satisfaction and exposure. This indicated that although students reported a high level of satisfaction with the Internet, this satisfaction does not lead to or originate from spending more time on the Net. Rather, satisfaction with the Internet is related to the level of importance that these students associate with the Internet in their lives. A sense of affinity, therefore, seems to be a prerequisite to satisfaction with the Internet, and/or to spending more time using it (Figure 11.1).

I examined three operational skills necessary for Internet use in this study. Those were typing speed, English-language proficiency, and Internet use proficiency. Unlike Chao (2001), who studied Internet use among Taiwanese students, most of the Arab respondents in our sample were quite proficient in terms

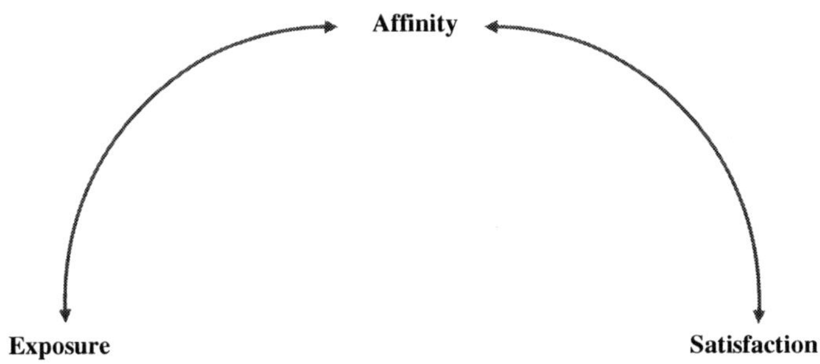

Figure 11.1. Relationship between Internet exposure, affinity, and satisfaction.

of operational skills. For example, while 69.8% (n = 326) of Chao's sample cited the "lack of English skills in reading and writing on the Internet" and 55.9% (n = 261) cited the "lack of adequate knowledge or skills to use the Internet" as obstacles to Internet use, most students in my sample (78.9%, n = 393) reported a high level of English-language proficiency, while another 19.9% (n = 99) reported a moderate proficiency level. As for Internet use proficiency, one-third of respondents reported being intermediate users (33.3%, n = 165), and two-thirds (66.7%, n = 331) reported being expert users. No respondents in our sample reported a low level of Internet use proficiency. Typing skills were not a major obstacle either, with slow typists (under 25 words per minute) constituting only 14.3% (n = 71) of the sample. The rest of the respondents reported being either moderate or fast typists. These high levels of proficiency, while maybe not reflective of the Egyptian or Arab population as a whole, are reflective of university-age students who have the economic means to afford to use the Internet. Most are educated in English-language schools, and are likely to own a personal computer. I do recognize, however, that lack of such operational skills, as well as cost, are potential obstacles to Internet adoption in the Arab world among the lower socioeconomic and educational classes.

The high English-language proficiency level reported by this sample may be one reason why they do not use much Arabic in their Internet experiences. This again is in sharp contrast to Chao (2001), who reported that the majority of Internet users in her Taiwanese student sample (97.6%, n = 487) used Chinese as their main language on the Internet. Moreover, the majority of the sites her respondents reported accessing on the Web were "within Taiwan" (97.4%, n = 486). Respondents in our sample reported using Arabic words spelled phonetically in English letters more than using Arabic characters (although both accounts are in relatively low percentages). This may be due to the lack of Arabic-enabled systems on most personal computers, as well as the fact that students in Egypt do not learn Arabic typing as part of their normal schooling, and are not required to submit typewritten documents in Arabic. On the other hand, AUC students (as well as computer and engineering students in other Egyptian and Arab universities) learn English typing either through a formal college class, or through experience since they are required to type their school assignments. Any other experiences students gain as everyday computer users help improve their English, rather than Arabic, typing skills.

The minimal use of the Arabic language by respondents is also a reflection of the problem of lack of Arabic content on the Internet. As of September 2004, Arabic-language content accounted for only 1.7% of what is available on the

Web (Global Reach, 2004). English-language content, in comparison, accounted for 35.2%, while Chinese accounted for 13.7%, and Spanish 9.0%.

The next variable I examined was type of Internet connection. Users of dial-up modem and users of fast access Internet connections did not seem to differ in their motives for using the Internet. However, they did differ in the time they spent on the Internet in an average week. Those who had fast access to the Internet spent more time on the network in an average week (22.63 hours) than those who accessed it through a dial-up modem (17.50 hours).

I also examined in this study how Internet use affects the amount of time dedicated to other media and other social activities among students. Over 70% of respondents said using the Internet has not affected their time to socialize with family, socialize with friends, or go to the movie theater. However, while the majority of respondents believe most of their media and social habits and behaviors have not been affected by their Internet use, some activities have been negatively affected. The medium that seems to have been affected the most is television, with 45% of respondents saying they now watch less television. Other activities that were negatively affected include reading newspapers (reported by 33% of respondents), reading magazines (30%), reading nonacademic books (29%), studying (28%), socializing with family (21%), and making phone calls (26%). On the other hand, 36% of respondents reported an increase in their listening to music on tapes or CDs, watching movies (18%), studying (15%), and socializing with friends (14%). It was concluded therefore that, with the exception of television, it does not seem that this study yielded clear-cut indications of the effect of Internet use on other media and social activities. (For more details, see Table 10.8 in the preceding chapter.) This does not depart from the studies that have tried previously to assess those effects, perhaps due in part to the difficulty of measuring such changes in behaviors. The findings of this study regarding the negative effect of Internet use on the amount of television viewing is in line with Choi (2001), who found that Korean respondents reported a decreased use of television with Internet use. It is also in line with Atkin et al. (1998), and Ferguson and Perse (2000), who reported that the Web might provide a functional alternative to television, especially in its use for diversion.

Finally, I tried to assess which factors or motives for use predicted Internet exposure among Arab students. Internet exposure in terms of total time spent on the Internet in an average week was significantly predicted by motives of social interaction and entertainment, as well as by type of Internet connection, and Internet affinity. As mentioned before, social interaction on the Internet is time consuming, and so it is logical to assume that this motive would positively predict exposure. Fast access to the Internet also makes it more convenient to spend more time on the Net, since the computer is logged on at all times for a flat rate fee. These results are in line with Papacharissi and A. Rubin (2000), who found that total Internet use was predicted by interpersonal utility (including a priori variables from the affection, inclusion, expression, social interaction, control, and surveillance categories).

E-mail use was predicted by type of Internet connection, surveillance motive, and age. Web use was predicted by entertainment motive, social interaction motive, type of Internet connection (implying that those with fast access connections were more likely to spend time using the Web), affinity, gender (implying that males are likely to spend more time on the Web than females), and age. Chat, instant messaging, and use of discussion boards was predicted by personal utility, age (negatively), social interaction motive, information seeking, and entertainment motive.

· 12 ·

THE POTENTIAL EFFECTS OF THE INTERNET IN EGYPT AND THE ARAB WORLD: A FUTURISTIC OUTLOOK

I tried to look at several issues in this book which I feel are of major importance when examining the diffusion, uses, and potential effects of the Internet in the Arab world. First, I looked at the current infrastructure and status of telecommunications and information technologies. Then I examined how governments deal with the Internet and the amount of freedom, or lack thereof, that these governments allow for the online flow of information. And finally, I presented the study I carried out on the uses of the Internet by Arab students.

It seems that the Internet, this wonderful information technology innovation that the West has created and embraced, still poses somewhat of a mystery to the Arab world and a challenge to the Arab governments, with the possible exception of a few countries.

Most Arab governments have been very cautious about approaching the Internet. For one thing, they are aware of the opportunities this medium offers to their countries for a better future and a more prosperous economy and people. On the other hand, they are wary of giving their people full freedom to access all that the Internet can offer. In the meantime, the slight percentage of Arab citizens whose education and socioeconomic status allow them to have access to the Internet have been trying to get the most out of it. Where their governments have imposed strict rules of blockage or censorship, they have attempted, and mostly succeeded, to outsmart these governments and circumvent censorship, even if through dialing an international ISP in a neighboring country. They have also used blogs to try to mobilize the masses, a behavior which some Arab governments find alarming. In this freedom-thirsty part of the world, Arab citizens long for any channel of communication that could help them gain access to more information or carry their own voices to each other and/or to the outside world.

I also believe the study presented in this book constitutes an important step in providing much needed, basic information about the use of the Internet by Arab citizens. The Internet has the potential to facilitate much change in this part of the world. The fact that this medium is primarily being used by students for information seeking and surveillance is an encouraging thought, since it is an indication that the audiences are aware of the potential of this new medium, particularly in countries where governments have controlled the information and media systems for the longest time. As an Egyptian, I believe the Internet can expose Arabs to new cultures, new patterns, new aspects of knowledge, and new ways of thinking. Of course this is precisely what some fear of the Internet, arguing it can also expose them to "immoral" or "inappropriate" material. Well, so can satellite television, and so can mobile phones, and so can computers. Two options present themselves for us Arabs in such situations: Either to tuck ourselves in and hide from the outside world, or to embrace modern technologies, and rely on the family and societal institutions to maintain morality and appropriate behavior. I firmly believe that the first choice would be in effect our own death sentence (aside from the fact that it is practically impossible). I believe that the potential of the Internet and information technologies to Arab countries is great, and should be utilized. Morality and "good behavior" are values that Arab families treasure very much, and they are values our Arab families can and do instill in their children, according to what Arab culture and religions subscribe to. Closing our doors and windows to the outside world by censoring and blocking Web sites is not a way to protect our children; it is a way to drown in ignorance and isolate ourselves from the wonders of science and knowledge and the road to freedom of speech and expression.

This study showed that the most salient interest for Arab students on the Internet is knowledge and information seeking. This should encourage Arabs to save no effort to provide Internet access to all, and increase Internet penetration in Arab countries. Parental guidance can and probably should be exercised, but only within the confines of the family. Governments, however, should not assume "parental responsibility" over their peoples, adults and children alike. In doing so, governments serve their own political interests, while hiding behind a mask of morality.

Social interaction on the Internet, although a new concept to the Arab world, has the potential to provide a basis for the creation of a more understanding and a more accepting people, a more robust civil society, as well as for presenting a more accurate image of Arabs to the outside world. Arabic chat rooms on the Internet range from economics to politics to cartoons to the latest in space shuttle technology. Such platforms provide a free space for Arabs, after years of government-owned and controlled media. So far, they seem to welcome it. After the attacks of September 11, 2001, for example, a heated debate was featured on several Arab discussion boards tackling every aspect of the attacks and what they meant for the United States and for the Arab world and the Muslim world. Arabs, Muslims and Christians alike, took to the Internet to explain to the world that terrorism is not Islamic, and has nothing to do with Islam. They explained the reasons behind the Arab person's frustrations with U.S. foreign policies, and presented a proud, informed illustration of what Islam is (Abdulla, 2005, in press). The main study featured in this book showed that the predictors of the social interaction motive indicated that it was still a habitual use for Arabs. However, as they become more familiar with it, and particularly as blogging sweeps the Internet world, Arabs are coming to realize the importance of such a tool for expressing themselves freely, and its potential for the creation of online virtual communities that could lead to a more democratic and more active civil society, and a more tolerant and informed world.

Jones (1995) spoke of the potential of the Internet to "create new opportunities for participatory democracy" (p. 26). Indeed, the Arab world has begun to ride on that wave. The past two years have witnessed a revival of NGOs and their role in society. Citizens are encouraged to participate as they actually see their efforts coming to fruition or feel a difference they make in society. Many of these NGOs use their Web site and/or e-mail as the main communication channel. Some of them have obtained licenses to serve as ISPs in an effort to raise funds for their different causes.

Several examples of NGOs are worth featuring. On the political front, *El Ghad*,[1] one of Egypt's main opposition parties, drew attention to itself by establishing an online radio station on its Web site in preparation for the 2005 presidential election. *Kifaya*,[2] an active Egyptian opposition movement that comprises people from all political fronts, mainly used the Internet to form and organize itself. *Shayfenkom*,[3] a political movement whose purpose is to monitor Egyptian elections, also mainly used their Web site to gain membership, draw volunteers, and to communicate their actions.

On the Egyptian social front, *Mossada*[4] uses e-mail to organize visits to orphanages, elderly homes, as well as to raise funds. Membership in the NGO stands at over 3500. Their projects include recycling, marriage loans, as well as donations and visits to hospitals, elderly homes, orphanages, and providing food and health caravans. *Resala*,[5] which started as a student movement, used its Web site to increase membership. The NGO now has 11 branch offices all over Egypt. It sponsors more than 40,000 needy families and 110 orphans, who have 951 volunteer big brothers and sisters. The NGO now has its own orphanage and hospital, which serves over 3000 patients a month for no charge. *Fathet Kheir*[6] or Volunteers for Development also uses the Web as its main communication channel. The NGO aims to help the needy, particularly women, through empowerment rather than charity. Their efforts include interest-free loans, informal training on different skills, literacy and educational classes, as well as regular health and awareness sessions for their clients.

Major efforts are needed though by Arab governments and IT interest groups if Internet use is to spread beyond the highly educated and affluent socioeconomic classes. Some of these efforts have started, exemplified by the free Internet initiative and the creation of technology clubs in Egypt, and numerous Internet cafes in Algeria, Tunisia, Egypt, Jordan, Morocco, Kuwait, and other countries. But other efforts need to be actively pursued. The most obvious of these efforts is providing much more Arabic content, and the necessary technical hardware and software to support it. The Arab society at large has to be educated about the new culture of the Internet, so that it would not be feared as some Western device that is out to spread immoral values or some sort of cultural imperialism. This study indicates that Internet users do not view the Internet as a threat to their cultural norms and identities. On the contrary, they see it as a tool that could help Arabs understand more about the rest of the world, and as an opportunity for Arabs to present their views and culture to the rest of the world.

There is also an argument that sometimes surfaces in the Arab world, stating that Arabs have more pressing priorities than the Internet, which stands to benefit only an affluent group of people. I have to note here that I strongly oppose this notion, as do many eminent societies (see, for example, UNDP,

2001) around the world, who argue for the Internet as a tool of national development and economic prosperity. However, even on the most basic level, I urge Arab citizens to consider the following example: Let's say both my porter and myself need to pay our phone bills. We can do that in two ways: either by going to the central phone office and stand in a long line to pay, or pay online through the Internet. I will obviously choose the more convenient way, I will pay online. My porter, however, is among the many who cannot use the Internet maybe because of illiteracy, lack of access, or any other reason. His only solution therefore is to go stand in the line and pay the bill. My access to the Internet in this case would still have helped him, because when I, and many others like me, pay through the Internet, he has to stand in a line that has 20 people instead of 50. Therefore we both win: I get to pay my bill in a most convenient way from the comfort of my home or office, and he gets to stand in a much shorter line than he would have otherwise. Of course there are other basic benefits to the phone company and the government here, including less crowded offices, less edgy public servants, more pleasant service, less paperwork being thrown around, and so on. The list is endless.

This simple example is typical of all Arab countries (and all countries worldwide) as well as of so many other situations. So the Internet, even if it may not appear to serve the less fortunate classes directly, will serve them indirectly at many levels, the most basic of which is to provide them with more space to get their work done and their services provided more efficiently.

This study provided a primary building block of knowledge in terms of assessing the uses and gratifications of the Internet among Arab students in Egypt. Future research on the Internet in the Arab world has a lot to accomplish. More specific uses of each of the functions of the Internet need to be assessed. Profiles of users at large are also needed: Who uses the Internet in each of the Arab countries? How do these users differ in terms of what they seek? What are their favorite activities or Web sites on the Internet? How do the psychological and social profiles of each user group differ, and how do those affect their Internet use experiences? An assessment of the effects of the Internet on individual users as well as on societies in general will also be needed in the near future, as well as an assessment of how gratifications influence such effects. This can help social science researchers figure out how the Internet can be used to contribute to a more robust civil society in the Arab world.

On the other hand, research on Internet avoidance is also needed. Why do those who avoid the Internet do that? Are the reasons mostly externally attributed (such as cost, lack of access, language barriers, lack of computer knowledge, etc.) or internally attributed (such as perceiving the Internet as a conflict to cultural norms, traditions, or religion) or a combination of both? Again, answering such questions would help enhance efforts to spread the Internet to the less affluent socioeconomic classes of the Arab world.

Concluding Remarks

There is no doubt that the Internet presents an open window onto the world. What we choose to look at or display is at our own discretion. And what we choose to look at or display, as well as what we choose to do with it, is what determines how it is likely to affect us. Never before has a medium provided such an easy access to information, bringing us a few steps closer to the potential of a true global village. Never before has it been this easy for a person or a people, no matter what their views, nationalities, religions, orientations, or cultures are, to share what they have to offer with each other and with the world. Arab governments, interest groups, IT professionals, and citizens should seize this opportunity to enable citizens of the Arab world to learn about others, and to let others learn about them. It would be futile for some of the more conservative Arab governments to waste their time trying to censor the contents of the Internet, or to exercise control over information or discussion boards emanating from or getting through to their countries. Arab officials are smart enough to know that, beyond a few minor nuisances, such techniques would just not work in the face of the advanced technologies that the Internet offers. When satellite broadcasting was introduced to the region, it became apparent how much more difficult it was going to be for Arab governments to censor that new transnational broadcasting technology compared to the traditional media that they owned, controlled, and had no difficulty censoring and shaping as they so pleased. With the advancement of technology, censorship became much more expensive and much less effective. The Internet is another step further up the anticensorable media scale. As Alterman (1998) pointed out, if governments wanted to censor the Internet, they would have to play "a game of catch-up" by closely monitoring the vast content of the Internet and blocking access to certain Web sites. They would be faced with the uphill challenge of just keeping up with what new content is uploaded on the network everyday. And even then, there are other ways for audiences to gain access to blocked sites, such as receiving the information by e-mail (encrypted if need be) from an account outside the borders of the censoring country, or as many do in the Arab countries that exercise censorship, access the Internet through a foreign ISP. Alterman suggested, and I agree, that potential censorship efforts by governments would not be effective, because usually those who want to evade the rules are much more technology savvy than those trying to implement them. If individuals want to block some sites from their homes, they should have access to free filtering software that they would then set up on their own computers to serve their own personalized needs.

The nature of the Internet as a mass medium is good news for the Arab people. Arabs should seize the opportunities of what the Internet has to offer. It could be their chance to enhance their lives on many fronts, not the least of which is cultivating a more tolerant and more active civic society. Efforts should concentrate on presenting more content of Arab origin and relevance, both in English and in Arabic, to the world. Egypt, as a historical leader of providing media content to the Arab world, should lead the way in this endeavor. The Internet could be used by Arabs to present a more truthful image of their culture and their religions to the Western world, a task that has become all the more important in light of world developments since the attacks of September 11, 2001. Such content, as well as online exchanges and friendships with people of different origins and cultures, could save Arabs a lot of unnecessary stereotyping that they face worldwide. Online friendships with Danish youth are a lot more beneficial to Arabs and Muslims than burning a Danish embassy. If Arabs ever complained about the negative portrayals of Arabs and Muslims on Western media, here is a good chance to rectify this image through a medium that readily penetrates the lives of millions of people all around the globe. As Jones (1995) put it, the information superhighway, brought about by the Internet, can do "what cement roads were unable to do, namely, connect us rather than atomize us, put us at the controls of a 'vehicle' and yet not detach us from the rest of the world" (p. 11). This is true for intercultural communication as much as for communication within the same community or culture.

The Internet is a medium that has the potential to reach millions the world over, and millions in the Arab world have the financial and educational capabilities to use it to their advantage and to the betterment of their societies. Arabs need to realize the medium's potentials, and Arab governments need to work on enhancing their IT infrastructure and capabilities, and on introducing Internet and computer education in all grade schools and in colleges and universities. Academic and applied research endeavors in the Arab world are needed every step of the way to guide this process. Only by realizing that the opportunities presented by the Internet far exceed any potential threats can Arabs become effective players in the global information system, and therefore use the medium to bridge the digital divide and bring more prosperity to their individual countries and the Arab region. Communication and friendship can go a long way for Arabs to rectify their image in the West, and share their values and their rich cultures with the rest of the world.

NOTES

Chapter 1

Parts of this chapter, chapter 2, and chapter 5 have been used in a paper for the *Global Media Journal—Mediterranean Edition*. The full citation is: Abdulla, R. (2006). An Overview of Media Developments in Egypt: Does the Internet make a difference? *Global Media Journal—Mediterranean Edition, 1*(1). Reprinted with permission.

1. These countries are: Algeria, Bahrain, Egypt, Iraq, Jordan, Kuwait, Lebanon, the Libyan Arab Jamahiriya, Morocco, Oman, the Occupied Palestinian Territories, Qatar, Saudi Arabia, Syrian Arab Republic, Sudan, Tunisia, the United Arab Emirates, and Yemen.
2. Adult illiteracy in the Arab world ranges from over 50% in countries such as Morocco and Yemen, to under 25% in richer countries such as Saudi Arabia, Kuwait, Jordan, and the United Arab Emirates (UAE). In Egypt, adult illiteracy is estimated at 44%. Overall adult illiteracy across the Arab world in 2002 amounted to 39% (UNDP, 2005).
3. This statistic, compiled in Sakr (2001) from several sources, included Israel, with an estimate of 69.4%. The statistic might therefore have inflated estimates of average satellite television access in those ten Arab countries.

Chapter 4

1. The Arab Human Development Report 2003 estimated the total number of Internet users in the Arab world in 2001 to have reached 4.2 million, a "considerable increase" from the 1 million users the report estimated in 2000 (UNDP & AFESD, 2002, 2003). However, with a large Internet user growth rate, most sources (Abdulla, 2001, 2005; Global Reach, 2004; Internet World Stats, 2006; ITU, 2001; MCIT, 2006), as well as individual country figures, indicate the number was around 21 million in September 2006.

Chapter 5

An abridged version of this chapter has been published in the Sage journal, *Global Media and Communication*. The full citation is: Abdulla, R. (2005). Taking the E-train: The development of the Internet in Egypt. *Global Media and Communication*, *1*(2), 149–165. Reprinted with permission.

1. At that time, private businesses had to go through an application process and justify their need for the service.
2. Telecom Egypt offers residential subscribers 1,600 free call units (of six minutes each) per year. After these units are used, the user is charged 10 P.T. (U.S. 1.7 cents) per six-minute unit of call time (ITU, 2001; MCIT, 2006).
3. *Al Gomhuria* is Arabic for "the republic."
4. *Al Ahram* is Arabic for "the pyramids."
5. *Sahafa* is Arabic for "journalism." The site can be accessed at www.sahafa.com.
6. Mr. Amr El Messidi was Marketing Manager of LINKdotNET at the time of the interview.
7. Little Horus can be accessed in Arabic at http://little-horus.com. The English version can be accessed at http://little-horus.com/en
8. *Otlob* is Arabic for "place an order."
9. With an unemployment crisis in the country, a lot of multilingual university graduates welcomed the establishment of these call centers, and rushed to apply to be part of the team of operators. Over 200,000 of them were employed by the centers.
10. http://www.idsc.gov.eg
11. Accessed though www.egypt.gov.eg or www.alhokoma.gov.eg. Al *Hokoma* is Arabic for "the government."
12. http://www.140online.com
13. For example, www.touregypt.net
14. www.tourism.egnet.net
15. www.eternalegypt.org.

Chapter 6

1. http://www.jannah.org/mamalist/
2. http://www.islamicsites.com/
3. http://www.islamicport.com/
4. http://www.islamicsites.20m.com/
5. The word "*fatawa*" is the Arabic plural of "*fatwa*."
6. http://www.maranao.com/index.php?option=com_wrapper&Itemid=40
7. http://www.islamicity.com and www.islam.org
8. http://www.sultan.org
9. http://www.islamweb.net
10. http://www.alazhar.org
11. http://www.jannah.org
12. http://www.islamonline.net
13. A Danish newspaper had in September 2005 published 12 cartoons of Prophet Mohammad (including one which featured the Prophet with a bomb-turban on his head), which

Muslims worldwide perceived as insulting and offensive. Muslim rage ignited all over the world and took all forms from letters to newspaper and magazine editors, to television productions, to worldwide demonstrations. Some extremists carried out acts of vendetta against Danish embassies in a few countries.

14. http://www.amrkhaled.net
15. http://www.arches.uga.edu/~godlas/
16. http://www.uga.edu/islam/jihad.html
17. http://www.uga.edu/islam/nineeleven.html
18. Yusuf Islam's Web site (http://www.yusufislam.org.uk) is itself an interesting source of information on Islam.
19. A minaret is a tall tower with a small balcony on top of a mosque, traditionally used as the place from which to call for prayers, so as to be heard as far as possible. It is a sign of a mosque, just like the cross is a sign of a church.
20. Muslims are required to pray five times a day (each prayer taking two to five minutes). They can pray at home or in a mosque, except for the Friday noon prayer, which men are required to carry out at a mosque in order to listen to a sermon by a Muslim sheikh. Praying at a mosque, although not required for all other prayers, is preferred and sought after by many Muslims.
21. A holy month in the Muslim calendar, when Muslims fast from sunrise to sunset every day. It is the month when the Holy Quran ascended onto Prophet Mohammad.
22. A three-day-feast, commemorating the end of Ramadan.
23. A four-day-feast, during which Muslims celebrate the hajj or pilgrimage.
24. The sayings of the Prophet that were authenticated by Muslim scholars, also referred to as Sunnah.

Chapter 7

1. In particular, they've reported using HTTPtunnel, Socks2HTTP, and CCProxy.
2. Human Rights Watch reported that the Egyptian government has blocked access to the Muslim Brotherhood site (http://www.ikhwanonline.com). However, the author has tried accessing the site many times from several different Internet connections, and at no time has she faced any difficulty in gaining access.

Chapter 8

1. Arabic for "Justice and Compassion."

Chapter 9

1. There does not seem to be an established acceptable response rate for Web-based or online surveys as yet, partly because of the difficulty of calculating one. Most studies using this methodology (see, for example, Choi, 2001) resort to posting a survey on a portal Web site and using the first quota of responses they receive within a given time (usually about 300).

Chapter 10

1. Five people did not mention their gender. Any other discrepancies in numbers that do not add up to a total of 502 are similarly due to missing data.

Chapter 12

1. Arabic for "tomorrow."
2. Arabic for "enough."
3. Arabic for "we can see you" or "we're watching you."
4. Arabic for "help."
5. Arabic for "message" and also "mission."
6. Arabic for "a good start."

REFERENCES

Abdel Raouf, M. (1998). *Egypt and the information society*. [In Arabic]. Cairo, Egypt: Dar Al Thaqafa.

Abdel Salam, N. (1998). The Egyptian and Arab electronic journalism experience: Reality and future prospects. [In Arabic]. *The Egyptian Journal for Media Research, 4*.

Abdulla, R. A. (1991). *The effect of CNN on the Egyptian news during the Gulf war*. Unpublished term paper, The American University in Cairo, Cairo, Egypt.

Abdulla, R. A. (1995, Summer). Welcome to the world of the Internet. *Brains Beyond Borders, 1*(1), 1, 3.

Abdulla, R. A. (1996). *The uses and gratifications of Music Television in Egypt: Why kids want their MTV*. Unpublished master's thesis, The American University in Cairo, Cairo, Egypt.

Abdulla, R. A. (2000). Why kids want their MTV: A cross-cultural perspective. In L. Gher & H. Amin (Eds.), *Civic discourse in the Middle East and digital age communication* (pp. 19–33). Greenwich, CT: Ablex.

Abdulla, R. A. (2001, November). *The Internet in the Middle East: An overview of developments and current status*. Paper presented at the National Communication Association Convention, Atlanta, GA.

Abdulla, R. A. (2003). The uses and gratifications of the Internet among students in Egypt. (Doctoral dissertation, University of Miami, 2003). *Dissertation Abstracts International*, DAI-A 65/01, 11.

Abdulla, R. A. (2004). Entertainment-education in the Middle East: Lessons from the Egyptian oral rehydration therapy campaign. In A. Singhal, M. Cody, E. Rogers, & M. Sabido (Eds.), *Entertainment-education and social change: History, research, and practice* (pp. 301–320). Mahwah, NJ: Lawrence Erlbaum Associates.

Abdulla, R. A. (2005). What they post: Arabic-language message boards after the September 11 attacks. In M. Salwen, B. Garrison, & P. Driscoll (Eds.), *Online news and the public* (pp. 279–302). Mahwah, NJ: Lawrence Erlbaum Associates.

Abdulla, R. A. (2006). Electronic transparency in the Arab world: How governments censor the Internet. [In Arabic]. *Al Hayat*.

Abdulla, R. A. (in press). Islam, jihad, and terrorism in post 9/11 Arabic discussion boards. *Journal of Computer-Mediated Communication.*

ACS (2002). The Academic Computing Services homepage. The American University in Cairo. Retrieved November 25, 2002 from http://www.aucegypt.edu/acs

Al Khuwailadi, M. (2001, March 30). Saudi Internet official notes 500 requests made daily to block sites. *Arab News.*

Alterman, J. (1998). *New media, new politics? From satellite television to the Internet in the Arab world* (Policy Paper No. 48). Washington, DC: The Washington Institute for Near East Policy.

Amin, H. (2000, April). *Arab media on the Internet: Competitive patterns and content issues.* Paper presented at the Broadcast Education Association, Las Vegas, NV.

Atkin, D. J., Jeffres, L. W., & Neuendorf, K. A. (1998, Fall). Understanding Internet adoption as telecommunications behavior. *Journal of Broadcasting & Electronic Media, 42*(4), 475–490.

AUC Office of Planning and Institutional Research. (2002). *AUC Profile 2002–2003* [Brochure]. Cairo: AUC Press.

Batelco (Bahrain Telecommunications Company) Web site. (2006). Retrieved January 6, 2006, from http://www.batelco.com.bh

Boyd, D. (1977, February). Egyptian radio: Tool of political and national development. *Journalism Monographs, 55*(3), 501–507, 539.

Boyd, D. (1998, August). *Saudi Arabia's international media strategy: Influence through multinational ownership.* Paper presented at the Association for Education in Journalism and Mass Communication, Baltimore.

Boyd, D. (1999). *Broadcasting in the Arab world: A survey of the electronic media in the Middle East* (3rd ed.). Ames, IA: Iowa State University Press.

Bunt, G. (2000). *Virtually Islamic: Computer-mediated communication and cyber Islamic environments.* Cardiff: University of Wales Press.

Bunt, G. (2003). *Islam in the digital age: E-Jihad, online fatwas and cyber Islamic environments.* London: Pluto Press.

Call Centers Egypt Web site. (2006). Retrieved January 3, 2006, from http://www.callcenters-egypt.com

Center for Development Communication (1989). *The Family House: A public health soap opera.* Proposal submitted to the Center for Communication Programs, Johns Hopkins University, Baltimore.

Chao, C. (2001). *Learning with the Internet: A quantitative and qualitative analysis of university students' Internet use and perceptions in Taiwan.* Unpublished doctoral dissertation, The Pennsylvania State University, PA.

Charney, T., & Greenberg, B. (2002). Uses and gratifications of the Internet. In C. A. Lin & D. J. Atkin (Eds.), *Communication technology and society: Audience adoption and uses* (pp. 379–407). Cresskill, NJ: Hampton Press, Inc.

Choi, Y. (2001). *Investigating Koreans' Internet use patterns and motivations, and exploring vulnerability of Internet dependency.* Unpublished doctoral dissertation, The University of Southern Mississippi, MS.

Clifford, P. (1995). Travelers on the Internet: A survey of Internet users. *Online, 19*(2), 29–34. Retrieved May 4, 2002, from http://www.onlineinc.com/onlinemag/OL1995/MarOL95/perry.html

Coffey, S., & Stipp, H. (1997). The interactions between computer and television usage. *Journal of Advertising Research, 37*(2), 61–67.

Cohen, J., & Cohen, P. (1983). *Applied multiple regression/correlation analysis for the behavioral sciences* (2nd ed.). Hillsdale, NJ: Lawrence Erlbaum Associates.

Conway, J. C., & Rubin, A. M. (1991). Psychological predictors of television viewing motivation. *Communication Research, 18*, 443–464.

Dabbagh Information Technology. (2001). *Arabic domain name boom*. Retrieved January 25, 2002, from http://www.dit.net/itnews/Article.asp?Article=1381

Dahel, R. (2001). *Telecommunications privatization in Arab countries: An overview*. Kuwait: Arab Planning Institute.

Diase, M. (1993, May). *Serials are mostly entertainment, but Egyptians like to learn from them*. Paper presented at the International Communication Association, Washington, DC.

Dillman, D. (2000). *Mail and Internet surveys: The tailored design method* (2nd ed.). New York: John Wiley & Sons, Inc.

Dubai Internet City Web site. (2006). Retrieved January 6, 2006, from http://www.dic.ae

Dyson, E. (1993). Interactivity means "active" participation. *Computerworld, 27*(50), 33–34.

Ebersole, S. (2000). Uses and gratifications of the Web among students. *Journal of Computer-Mediated Communication*, 6(1), 1–20. Retrieved May 4, 2002, from http://www.ascusc.org/jcmc/vol6/issue1/ebersole.html

Egypt Information Highway. (2001). Retrieved January 25, 2002, from http://www.highway.idsc.gov.eg

Egypt Smart Village Web site. (2006). Retrieved January 8, 2006, from http://www.smart-villages.com

Egyptian Government Web site. (2006). Retrieved January 24, 2006, from http://www.egypt.gov.eg

Egyptian Universities Network Web site. (2001). Retrieved January 25, 2002, from http://www.frcu.eun.eg

Ein-Dor, P., Goodman, S., & Wolcott, P. (1999). *The global diffusion of the Internet project: The Hashemite Kingdom of Jordan*. Retrieved November 14, 2004, from http://mosaic.unomaha.edu/Jordan_1999.pdf

El Hoteiby, A. (1999). *Online journalism publishing in the Arab world: Performance and future impact*. Unpublished master's thesis, The American University in Cairo, Cairo, Egypt.

El Kashef, I. (2005). Islam dot com. *Al Ahram Weekly*. October 13–19. Retrieved February 24, 2006, from http://weekly.ahram.org.eg/2005/764/fo2.htm

Elewa, H. (1998). *The role of the Arab satellite stations in changing types of social behavior in a sample of the Egyptian public: An analytical field study*. Unpublished doctoral dissertation, Zaqazeeq University, Sharkiya, Egypt.

Elton, M., & Carey, J. (1983). Computerizing information: Consumer reactions to teletext. *Journal of Communication*, 33(1), 162–173.

Esmat, A. (2002, December 15). The digital divide threatens the Arab world. [In Arabic]. *Al Ahram*, 27.

Etisalat (Emirates Telecommunications Corporation) Web site. (2006). Retrieved January 6, 2006, from http://www.etisalat.co.ae

Ferguson, D., & Perse, E. M. (2000, Spring). The World Wide Web as a functional alternative to television. *Journal of Broadcasting & Electronic Media, 44*(2), 155–174.

Finn, S., & Gorr, M. (1988). Social isolation and social support as correlates of television viewing motivations. *Communication Research, 15*, 135–158.

Flaherty, L. M., Pearce, K. J., & Rubin, R. B. (1998, Summer). Internet and face-to-face communication: Not functional alternatives. *Communication Quarterly, 46*(3), 250–268.

Franda, M. (2002). *Launching into cyberspace: Internet development and politics in five world regions*. London: Lynne Rienner.

Global Reach. (2004). *Global Internet statistics*. Retrieved January 6, 2006, from http://global-reach.biz/globstats/index.php3

Gorsuch, R. L. (1983). *Factor analysis*. Hillsdale, NJ: Lawrence Erlbaum Associates.

Greenberg, B. S. (1974). Gratifications of television viewing and their correlates for British children. In J. G. Blumler & E. Katz (Eds.), *The uses of mass communication: Current perspectives on gratifications research* (pp. 71–92). Beverly Hills, CA: Sage.

Hamdy, N. (2001). *A profile of the Egyptian Internet user*. Presented to the Broadcast Education Association. Las Vegas.

Harman, H. (1976). *Modern factor analysis* (3rd ed). Chicago: University of Chicago Press.

Harris, R. (2001). *A primer of multivariate statistics* (3rd ed.). Mahwah, NJ: Lawrence Erlbaum Associates.

Hassan, J. (2002). Most Islamic websites are in Arabic: Survey. *Arab News*. Retrieved January 6, 2006, from http://www.arabnews.com/?page=1&section=0&article=20030&d=1&m=11&y=2002

Hassan, W. (1996). *The information superhighway: Prospects and concerns for the future of Egyptian mass media*. Unpublished master's thesis, The American University in Cairo, Cairo, Egypt.

Hatef Telecom Web site. (2006). Retrieved January 6, 2006, from http://www.hatiftelecom.com/news/oman.html

Human Rights Watch. (1999). *The Internet in the Mideast and North Africa: Free expression and censorship*. Retrieved November 14, 2004, from http://hrw.org/advocacy/internet/mena.htm

Human Rights Watch. (2005). *False freedom: Online censorship in the Middle East and North Africa*. Retrieved January 3, 2006, from http://hrw.org/reports/2005/mena1105/mena1105webw-cover.pdf

Human Rights Watch. (2006). *Jordan: Editor prosecuted for posting articles by MPs*. Retrieved January 15, 2006, from http://hrw.org/english/docs/2006/01/26/jordan12537.htm

Ibrahim, M. (1999). *The use of the Internet by the Egyptian press and its impact on journalistic performance*. [In Arabic]. Paper presented to the 5th Scientific Conference of the Media Faculty, Cairo University, Cairo, Egypt.

Information Technology Industry Development Agency Web site. (2006) Retrieved January 6, 2006, from http://www.itida.gov.eg

International Telecommunication Union. (1999). *Challenges to the network: Internet for development*. Retrieved January 25, 2002, from http://www.itu.int/ti/publications/Inet_99/chal_exsum.pdf

International Telecommunication Union. (2001). *Internet on the Nile: Egypt case study*. Retrieved November 1, 2002, from http://www.itu.int/itudoc/gs/promo/bdt/cast_int/80621.pdf

Internet Society Web site. (2006) Retrieved January 6, 2006, from http://www.isoc.org

Internet Society of Egypt. (2001). *Internet history in Egypt*. Retrieved January 25, 2002, from http://www.ise.org.eg/history.htm

Internet Systems Consortium, Inc. Web site. (2006) Retrieved January 6, 2006, from http://www.isc.org

Internet World Stats Web site. (2006) Retrieved September 25, 2006, from http://www.internetworldstats.com

Jarrah, F. (2000). *Number of Internet users in Arab countries edges towards two million.* Retrieved September 22, 2001, from the Dabbagh Information Technology Web site: http://www.dit.net/itnews/me_internet/users.html

Jeffres, L., & Atkin, D. (1996, Summer). Predicting use of technologies for communication and consumer needs. *Journal of Broadcasting & Electronic Media, 40*(3), 318–330.

Jones, S. (1995). Understanding community in the information age. In S. Jones (Ed.), *Cybersociety: Computer-mediated communication and community* (pp. 10–35). Thousand Oaks, CA: Sage.

Jones, S. (1997). The Internet and its social landscape. In S. Jones (Ed.), *Virtual culture: Indentity and communication in cybersociety* (pp. 7–35). Thousand Oaks, CA: Sage.

Jordan Telecommunication Company Web site. (2006). Retrieved January 25, 2006, from http://www.jordantelecom.jo

Kaiser, H. F. (1960). The application of electronic computers to factor analysis. *Educational and Psychological Measurement, 20,* 141–151.

Kaiser, H. F. (1970). A second generation Little Jiffy. *Psychometrika, 35,* 401–415.

Kaiser, H. F. (1974). Little Jiffy, Mark IV. *Educational and Psychological Measurement, 34,* 111–117.

Kandell, J. (1998). Internet addiction on campus: The vulnerability of the college student. *Cyberpsychology and Behavior, 1,* 11–18.

Katz, E., Blumler, J. G, & Gurevitch. M. (1974). Utilization of mass communication by the individual. In J. G. Blumler & E. Katz (Eds.), *The uses of mass communication: Current perspectives on gratifications research* (pp. 19–32). Beverly Hills, CA: Sage.

Katz, E., & Foulkes, D. (1962). On the use of the mass media as "Escape": Clarification of a concept. *Public Opinion Quarterly, 26,* 377–388.

Katz, J., & Aspden, P. (1997a). Motives, hurdles, and dropouts: Who is on and off the Internet and why. *Communications of the ACM, 40*(4), 97–102.

Katz, J., & Aspden, P. (1997b). A nation of strangers. *Communications of the ACM, 40*(12), 81–86.

Katz, J., & Aspden, P. (1997c). Motivations for and barriers to Internet usage: Results of a national public opinion survey. *Internet Research: Electronic Networking Applications and Policy, 7*(3), 170–188.

Kaye, B., & Johnson, T. (2002). Online and in the know: Uses and gratifications of the Web for political information. *Journal of Broadcasting and Electronic Media, 46*(1), 54–71.

Kim, J. & Mueller, C. (1978a). *Factor analysis: Statistical methods and practical issues.* Thousand Oaks, CA: Sage Publications, Quantitative Applications in the Social Sciences Series, No. 14.

Kim, J., & Mueller, C. (1978b). *Introduction to factor analysis: What it is and how to do it.* Thousand Oaks, CA: Sage Publications, Quantitative Applications in the Social Sciences Series, No. 13.

Kraut, R., Lundmark, V., Patterson, M., Kiesler, S., Mukopadhyay, T., & Scherlis, W. (1998). Internet paradox: A social technology that reduces social involvement and psychological well-being? *American Psychologist, 53*(9), 1017–1031.

Kraut, R., Mukhopadhyay, T., Szczypula, J., Kiesler, S. & Scherlis, W. (1998, April). Communication and information: Alternative uses of the Internet in households. In *Proceedings of the CHI 98* (pp. 368–383). New York: ACM.

Kruskal, J., & Wish, M. (1978). *Multidimensional scaling.* Thousand Oaks, CA: Sage Publications, Quantitative Applications in the Social Sciences Series, No. 11.

Lebanese Ministry of Telecommunications Web site. (2006). Retrieved January 6, 2006, from http://www.mpt.gov.lb

Leung, L., & Wei, R. (2000, Summer). More than just talk on the move: Uses and gratifications of the cellular phone. *Journalism & Mass Communication Quarterly, 77*(2), 308–320.

Lin, C. A. (1992, Spring). The functions of the VCR in the home leisure environment. *Journal of Broadcasting & Electronic Media, 36*(3), 345–351.

Lin, C. A. (1993a). Adolescent viewing and gratifications in a new media environment. *Mass Comm Review, 20*(1/2), 39–50.

Lin, C. A. (1993b). Modeling the gratification-seeking process of television viewing. *Human Communication Research, 20*(2), 224–244.

Lin, C. A. (1994, Winter). Exploring the role of VCR use in the emerging home entertainment culture. *Journalism Quarterly, 70*(4), 833–842.

Lin, C. A. (1999). Predicting online service adoption likelihood among potential subscribers: A motivational approach. *Journal of Advertising Research*, 39(4), 79–89.

Lin, C. A. (2001, Winter). Audience attributes, media supplementation, and likely online service adoption. *Mass Communication & Society*, 4(1), 19–38.

LINKdotNET Web site. (2006). Retrieved January, 8, 2006, from http://www.link.net

Little Horus Web site. (2006) Retrieved January, 8, 2006, from http://little-horus.com/en/Default.aspx

Maignan, I., & Lukas, B. (1997). The nature and social uses of the Internet: A qualitative investigation. *Journal of Consumer Affairs, 31*(2), 346–371.

Marghalani, K., Palmgreen, P., & Boyd, D. (1998). The utilization of direct satellite broadcasting (DBS) in Saudi Arabia. *Journal of Broadcasting and Electronic Media, 42*(3), 297–314.

Masmoudi, M. (1997). *The gateways of the new age of communication.* [In Arabic]. Tunis, Tunisia: Metamedia.

McLuhan, M. (1962). *The Guttenburg galaxy*. Toronto: University of Toronto Press.

McLuhan, M. (1964). *Understanding media: The extensions of man*. New York: McGraw-Hill.

Media Monitor. (2005, June 20). *Egypt: Television reinforces position as dominant medium*. Retrieved February 1, 2006, from http://news.monstersandcritics.com/mediamonitor/article_1017066.php/Egypt_Television_reinforces_position_as_dominant_medium

Media Production City: A colossal media complex. (n.d.). Retrieved October 27, 2002, from http://www.touregypt.net/mpc.htm

Mesbah, H. M. (1991). *Uses and gratifications of television viewing among Egyptian adults*. Unpublished master's thesis, The American University in Cairo, Cairo, Egypt.

Ministry of Communications and Information Technology Yearbook. (2004). Cairo: MCIT.

Ministry of Communications and Information Technology Web site. (2006). Retrieved February 6, 2006, from www.mcit.gov.eg

Ministry of Health and Population Web site. (2006). Retrieved January 24, 2006, from http://www.mohp.gov.eg/

MobiNil Web site. (2006). Retrieved February 6, 2006, from http://www.mobinil.com

Moon, D. (2002). *Islam on the Internet. Part III: The debate over online Muslim ideology*. Retrieved March 20, 2006 from the National Public Radio Web site: http://www.npr.org/programs/watc/cyberislam/ideology.html

Morris, M., & Ogan C. (1996). The Internet as mass medium. *Journal of Communication*, 46(1), 39–50.

Nagib, M. (2000, May). *Arabization of the Internet: Assessment and impact.* Unpublished master's thesis, The American University in Cairo, Cairo, Egypt.

National Telecommunication Regulatory Authority Web site. (2006). Retrieved February 6, 2006, from http://www.tra.gov.eg

Nie, N., & Erbring, L. (2000). *Internet and society: A preliminary report.* Retrieved February 12, 2003, from the Stanford Institute for the Quantitative Study of Society Web site: http://www.stanford.edu/group/siqss/Press_Release/Preliminary_Report.pdf

Norusis, M. (2002). *SPSS 11.0 guide to data analysis.* New York: Prentice Hall.

Open Net Initiative. (2004). *Internet filtering in Saudi Arabic in 2004.* Retreived January 3, 2006, from http://www.opennetinitiative.net/studies/saudi

Palmgreen, P. (1984). Uses and gratifications: A theoretical perspective. *Communication Yearbook, 8,* 20–55.

Palmgreen, P., & Rayburn, J.D., II (1979). Uses and gratifications and exposure to public television: A discrepancy approach. *Communication Research,* 6, 155–180.

Palmgreen, P., & Rayburn, J. D., II (1982). Gratifications sought and media exposure: An expectancy value model. *Communication Research,* 9, 561–580.

Palmgreen, P., & Rayburn, J. D., II (1985). A comparison of gratification models of media satisfaction. *Communication Monographs, 52,* 334–346.

Palmgren, M. (2001). Surfing at school: The buzz in Egypt is about educating those who shall use computers in the Internet age. *Egypt Today.* [Online version]. Retrieved November 1, 2002, from http://www.egypttoday.com/et200001/wnonli.htm

Papacharissi, Z., & Rubin, A. (2000). Predictors of Internet use. *Journal of Broadcasting and Electronic Media, 44*(2), 175–196.

Perry, T., Perry, L., & Hosack-Curlin, K. (1998). Internet use by university students: An interdisciplinary study on three campuses. *Internet Research: Electronic Networking Applications and Policy,* 8, 136–141.

Perse, E. M. (2001). *Media effects and society.* Mahwah, NJ: Lawrence Erlbaum Associates.

Perse, E. M., & Dunn, D. (1998). The utility of home computers and media use: Implications on multimedia and connectivity. *Journal of Broadcasting & Electronic Media, 42,* 435–456.

Perse, E. M., & Ferguson, D. (2000, Fall). The benefits and costs of Web surfing. *Communication Quarterly,* 48(4), 343–359.

Piecowye, J. (2003). *Habitus* in transition? CMC use and impacts among young women in the United Arab Emirates. *Journal of Computer-Mediated Communication,* 8(2). Retrieved November 6, 2004, from http://www.ascusc.org/jcmc/vol8/issue2/piecowye.html

Privacy International & the GreenNet Educational Trust. (2003). *Silenced: An international report on censorship and control of the Internet.* Retrieved November 14, 2004, from http://www.privacyinternational.org

Q-Tel (Qatari Public Telecommunication Corporation) Web site. (2006). Retrieved January 25, 2006, from http://www.qtel.com.qa

Qulaini, S. (1997). The effects of watching satellite television on social trends among Egyptian youth. *The Humanitarian Sciences and Arts Journals, 25,* 71–144.

Rayburn, J. D., II. (1996). Uses and gratifications. In M. Salwen & D. Stacks (Eds.), *An integrated approach to communication theory and research* (pp. 145–163). Mahwah, NJ: Lawrence Erlbaum Associates.

Read, B. (2001). A Georgia professor's Web site on Islam attracts new attention. *The Chronicle of Higher Education*, October 31. Retrieved March 14, 2006, from http://chronicle.com/free/2001/10/2001103101t.htm

Reporters Sans Frontieres. (2004). *Internet Under Surveillance 2004*. Retrieved November 14, 2005, from http://www.rsf.org

Reporters Sans Frontieres Web site. (2006). Retrieved September 15, 2006, from http://www.rsf.org

RITSEC. (2001). *Egypt Information Highway Project*. The Regional Information Technology and Software Engineering Center. Retrieved July 17, 2002, from http://www.ritsec.com.eg

Rubin, A. M. (1977). Television usage, attitudes and viewing behaviors of children and adolescents. *Journal of Broadcasting, 21*(3), 355–369.

Rubin, A. M. (1979). Television use by children and adolescents. *Human Communication Research*, 5(2), 109–120.

Rubin, A. M. (1981). An examination of television viewing motivations. *Communication Research*, 8, 141–165.

Rubin, A. M. (1983). Television uses and gratifications: The interactions of viewing patterns and motivations. *Journal of Broadcasting, 27*(1), 37–51.

Rubin, A. M. (1984). Ritualized and instrumental television viewing. *Journal of Broadcasting, 34*(3), 67–77.

Rubin, A. M. (1985a). Uses and gratifications. In J. Dominick & J. Fletcher (Eds.), *Broadcasting research methods* (pp. 193–209). Boston: Allen and Bacon.

Rubin, A. M. (1985b). Uses of daytime television soap operas by college students. *Journal of Broadcasting & Electronic Media, 29*(3), 241–256.

Rubin, A. M. (1993). Audience activity and media use. *Communication Monographs*, 60(1), 98–105.

Rubin, A. M. (1994). Media uses and effects: A uses-and-gratifications perspective. In J. Bryant & D. Zillmann (Eds.), *Media effects: Advances in theory and research* (pp. 417–436). Hillsdale, NJ: Lawrence Erlbaum Associates.

Rubin, A. M., & Perse, E. M. (1987a). Audience activity and soap opera involvement: A uses and effects investigation. *Human Communication Research*, 14(2), 246–268.

Rubin, A. M., & Perse, E. M. (1987b). Audience activity and television news gratifications. *Communication Research*, 13(1), 58–84.

Rubin, A. M., & Rubin, R. B. (1982). Contextual age and television use. *Communication Research*, 8(3), 228–244.

Rubin, A. M., & Rubin, R. B. (1985). Interface of personal and mediated communication: A research agenda. *Critical Studies in Mass Communication*, 2, 36–53.

Rubin, R. B., Palmgreen, P., & Sypher, H. E. (Eds). (1994). *Communication research measures: A sourcebook*. New York: Guilford.

Ruggiero, T. (2000). Uses and gratifications theory in the 21st century. *Mass Communication and Society*, 3(1), 3–37.

Sakr, N. (2001). *Satellite realms: Transnational television, globalization and the Middle East*. London: I.B.Tauris Publishers.

Saudi Internet Services Unit [ISU] Web site. (2004). Retrieved January 6, 2006, from http://www.isu.net.sa/index.html

Saudi Telecom Web site. (2006). Retrieved January 6, 2006, from http://www.stc.com.sa

Scherer, K. (1996). College life on-line: Healthy and unhealthy Internet use. *Journal of College Student Development, 38*, 655–665.

School Net Africa. (2003). *Egypt: 21st Century Kids Club Project*. Retrieved January 6, 2006, from http://www.schoolnetafrica.net/index.php?id=434

Schramm, W., Lyle, J., & Parker, E. (1961). *Television in the lives of our children*. Stanford: Stanford University Press.

Shaheen, H. (2001). *The Egyptian public and viewing Arab satellite channels: An analytical field study*. [In Arabic]. Unpublished doctoral dissertation, Cairo University, Cairo, Egypt.

SPSS for Windows. (2002). Rel. 11.5.0. Chicago: SPSS Inc.

Stempel, G. H. III, Hargrove, T., & Bernt, J. P. (2000). Relation of growth of use of the Internet to changes in media use from 1995 to 1999. *Journalism & Mass Communication Quarterly, 77*(1), 71–-79.

Sudatel. (2006). The Sudanese Telecommunications Company Web page. Retrieved January 6, 2006, from http://www.sudatel.net

Swanson, D. (1987). Gratification seeking, media exposure, and audience interpretations. *Journal of Broadcasting & Electronic Media, 31*(3), 237–254.

Technology Access Community Centers Web site. (2002). Retrieved November 3, 2002, from www.tacc.egnet.net

Thuraya Web site. (2006). Retrieved January 6, 2006, from http://www.thuraya.com

United Nations Development Programme. (2001). *Human development report 2001: Making new technologies work for human development*. New York: Oxford University Press.

United Nations Development Programme. (2002). *Human development report 2002: Deepening democracy in a fragmented world*. New York: Oxford University Press.

United Nations Development Programme. (2003). *Human development report 2003: Millennium development goals: A compact among nations to end human poverty*. New York: Oxford University Press.

United Nations Development Programme. (2004). *Human development report 2004: Cultural liberty in today's diverse world*. New York: Oxford University Press.

United Nations Development Programme. (2005). *Human development report 2005: International cooperation at a crossroads: Aid, trade and security in an unequal world*. New York: Oxford University Press.

United Nations Development Programme, & Arab Fund for Economic and Social Development. (2002). *Arab human development report 2002: Creating opportunities for future generations*. Amman, Jordan: UNDP Regional Bureau for Arab States.

United Nations Development Programme, & Arab Fund for Economic and Social Development. (2003). *Arab human development report 2003: Building a knowledge society*. Amman, Jordan: UNDP Regional Bureau for Arab States.

United Nations Development Programme, & Arab Fund for Economic and Social Development. (2004). *Arab human development report 2004: Freedom and good governance*. Amman, Jordan: UNDP Regional Bureau for Arab States.

United Nations Development Programme, & Egypt's Institute of National Planning. (2003). *Egypt human development report 2003: Local participatory development*. Cairo, Egypt: Al Ahram Commercial Press.

United Nations Development Programme, & Egypt's Institute of National Planning. (2004). *Egypt human development report 2004: Choosing decentralization for good governance*. Cairo, Egypt: Al Ahram Commercial Press.

United Nations Development Programme, & Egypt's Institute of National Planning. (2005). *Egypt human development report 2005: Choosing our future: Towards a new social contract*. Cairo, Egypt: Al Ahram Commercial Press.

United States Information Service. (2000). *IT sector and USAID assistance backgrounder*. Retrieved January 25, 2002, from http://www.usis.egnet.net/usis/bg5–24.htm

US-Egypt President's Council. (2002). *Egypt information and communications technologies dashboard: Quarterly update on key Egyptian information technology issues*. Retrieved November 5, 2002, from http://www.us-egypt.org/digitaldashseptember2002.html

Vodafone Egypt Web site. (2006). Retrieved February 6, 2006, from http://www.vodafone.com.eg

Weaver, A. (1998). Net worth. *Working Woman, 23*(1), 20.

Weiser, E. (2000). *The functions of Internet and their social, psychological, and interpersonal consequences*. Unpublished doctoral dissertation, Texas Tech University, TX.

Wells, A. (1990). Popular music: Emotional use and management. *Journal of Popular Culture, 24*(1), 105–117.

Wheeler, D. L. (1998). Global culture or culture clash: New information technologies in the Islamic world—A view from Kuwait. *Communication Research, 25*, 359–376.

Wheeler, D. L. (2003). The Internet and youth subculture in Kuwait. *Journal of Computer-Mediated Communication*, 8(2). Retrieved November 6, 2004, from http://www.ascusc.org/jcmc/vol8/issue2/wheeler.html

Wheeler, D. L. (2004). Blessings and curses: Women and the Internet revolution in the Arab world. In N. Sakr (Ed.), *Women and media in the Middle East: Power through self-expression* (pp. 138–161). London: I.B. Tauris.

Williams, F. (1992). *Reasoning with statistics: How to read quantitative research* (4th ed.). Fort Worth: Harcourt Brace Jovanovich College Publishers.

Williams, F., Phillips, A., & Lum, P. (1985). Gratifications associated with new communication technologies. In K. Rosengren, L. Wenner, & P. Palmgreen (Eds.), *Media gratifications research: Current perspectives* (pp. 241–252). Beverly Hills, CA: Sage.

Williams, F., Rice, R., & Rogers, E. (1988). *Research methods and the new media*. New York: Free Press.

World Economic Forum (2004). *The global information technology report 2003–2004. Towards an equitable information society*. New York: Oxford University Press.

World Economic Forum (2005). *The global information technology report 2004–2005. Efficiency in an increasingly connected world*. New York: Oxford University Press.

Youssef, H. (2001). *Arab issues as presented in the news of CNN and Euronews: A comparative study*. [In Arabic]. Unpublished doctoral dissertation, Cairo University, Cairo, Egypt.

Zitoun, H., & Sghaier, L. (2000). *How to adapt electronic commerce to the needs of a developing country: The Tunisian case*. Paper presented to the Internet Society. Retrieved November 14, 2004, from http://www.isoc.org/inet2000/cdproceedings/7c/7c_4.htm

INDEX

2502911874

Digital Formations

General Editor: *Steve Jones*

Digital Formations is an essential source for critical, high-quality books on digital technologies and modern life. Volumes in the series break new ground by emphasizing multiple methodological and theoretical approaches to deeply probe the formation and reformation of lived experience as it is refracted through digital interaction. **Digital Formations** pushes forward our understanding of the intersections—and corresponding implications—between the digital technologies and everyday life. The series emphasizes critical studies in the context of emergent and existing digital technologies.

Other recent titles include:

Leslie Shade
Gender and Community in the Social Construction of the Internet

John T. Waisanen
Thinking Geometrically

Mia Consalvo & Susanna Paasonen
Women and Everyday Uses of the Internet

Dennis Waskul
Self-Games and Body-Play

David Myers
The Nature of Computer Games

Robert Hassan
The Chronoscopic Society

M. Johns, S. Chen, & G. Hall
Online Social Research

C. Kaha Waite
Mediation and the Communication Matrix

Jenny Sunden
Material Virtualities

Helen Nissenbaum & Monroe Price
Academy and the Internet

To order other books in this series please contact our Customer Service Department:
(800) 770-LANG (within the US)
(212) 647-7706 (outside the US)
(212) 647-7707 FAX
To find out more about the series or browse a full list of titles, please visit our website:
WWW.PETERLANG.COM